George Egbert Fisher, Issac Joachim Schwatt, Heinrich Durège

Elements of the Theory of Functions of a Complex Variable

With Especial Reference to the Methods of Riemann

George Egbert Fisher, Issac Joachim Schwatt, Heinrich Durège

Elements of the Theory of Functions of a Complex Variable
With Especial Reference to the Methods of Riemann

ISBN/EAN: 9783337277062

Printed in Europe, USA, Canada, Australia, Japan

Cover: Foto ©berggeist007 / pixelio.de

More available books at **www.hansebooks.com**

ELEMENTS

OF THE

THEORY OF FUNCTIONS

OF A

COMPLEX VARIABLE

WITH

ESPECIAL REFERENCE TO THE METHODS OF RIEMANN

BY

DR. H. DUREGE

LATE PROFESSOR IN THE UNIVERSITY OF PRAGUE

AUTHORIZED TRANSLATION FROM THE FOURTH GERMAN EDITION

BY

GEORGE EGBERT FISHER, M.A., Ph.D.
ASSISTANT PROFESSOR OF MATHEMATICS IN THE UNIVERSITY OF PENNSYLVANIA

AND

ISAAC J. SCHWATT, Ph.D.
INSTRUCTOR IN MATHEMATICS IN THE UNIVERSITY OF PENNSYLVANIA

PHILADELPHIA
G. E. FISHER AND I. J. SCHWATT
1896

TRANSLATORS' NOTE.

WE desire to express our indebtedness to Professor James McMahon, Cornell University, for his kindness in reading the translation, and for his valuable suggestions. Also to Messrs. J. S. Cushing & Co., Norwood, Mass., for the typographical excellence of the book.

G. E. F.
I. J. S.

UNIVERSITY OF PENNSYLVANIA,
PHILADELPHIA, December, 1895.

FROM THE AUTHOR'S PREFACE TO THE THIRD EDITION.

NUMEROUS additions have been made to Section IX., which treats of multiply connected surfaces. If Riemann's fundamental proposition on these surfaces be enunciated in such a form that merely simply connected pieces are formed by both modes of resolution, — as is ordinarily, and was also in § 49, the case, — then it must be supplemented for further applications. Such supplementary matter was given in § 52 in the classification of surfaces, and in § 53, V. But if we express the fundamental proposition in the form in which Riemann originally established it, in which merely simply connected pieces are formed by only one mode of resolution, while the pieces resulting from the other mode of resolution may or may not be simply connected, then all difficulties are obviated, and the conclusions follow immediately without requiring further expedients.

This was shown in a supplementary note at the end of the book.

AUTHOR'S PREFACE TO THE FOURTH EDITION.

In the present new edition only slight changes are made, consisting of brief additions, more numerous examples, different modes of expression, and the like.

In reference to the above extract from the preface of the preceding edition, I have asked myself the question, whether I should not from the beginning adopt the original Riemann enunciation of the fundamental proposition instead of that which is given in § 49. Nevertheless, I have finally adhered to the previous arrangement, because I think that in this way the difference between the two enunciations is made more prominent, and the advantages of the Riemann enunciation are more distinctly emphasized.

H. DURÈGE.

PRAGUE, April, 1893.

CONTENTS.

	PAGE
INTRODUCTION	1

SECTION I.

GEOMETRIC REPRESENTATION OF IMAGINARY QUANTITIES.

§ 1. A complex quantity $x + iy$ or $r(\cos\phi + i\sin\phi)$ is represented by a point in a plane, of which x and y are the rectangular, r and ϕ the polar co-ordinates. A straight line is determined in length and direction by a complex quantity. Direction-coefficient 12
§ 2. Construction of the first four algebraic operations 15
    1. Addition 15
    2. Subtraction. Transference of the origin 16
    3. Multiplication 18
    4. Division. Applications 20
§ 3. Complex variables. They can describe different paths between two points. Direction of increasing angles 22

SECTION II.

FUNCTIONS OF A COMPLEX VARIABLE IN GENERAL.

§ 4. The correspondence of the values of the variable and of the function is the most essential characteristic of a function 25
§ 5. Conditions under which $w = u + iv$ is a function of $z = x + iy$ 27
§ 6. The derivative $\frac{dw}{dz}$ is independent of dz 29
§ 7. If w be a function of z, the system of w-points is similar in its infinitesimal elements to the system of z-points. Conformal representation. Conformation 33

SECTION III.

MULTIFORM FUNCTIONS.

		PAGE
§ 8.	The value of a multiform function depends upon the path described by the variable. Branch-points	37
§ 9.	Two paths lead to different values of a function only when they enclose a branch-point	43
§ 10.	Examples:	
	(1) $\sqrt{z}$, (2) $(z-1)\sqrt{z}$, (3) $\sqrt[3]{\dfrac{z-a}{z-b}}$, (4) $\sqrt[3]{\dfrac{z-a}{z-b}}+\sqrt{z-c}$.	
	Cyclical interchange of function-values	47
§ 11.	Introduction of Riemann surfaces which cover the plane n-fold. Branch-cuts	55
§ 12.	Proof that this mode of representation of n-valued functions is conformal	61
§ 13.	Continuous and discontinuous passage over the branch-cuts. Simple branch-points and winding-points of higher orders	68
§ 14.	The infinite value of z. Surfaces closed at infinity (Riemann spherical surfaces). Proof that the formation of a Riemann spherical surface for a given algebraic function is always possible. Examples of different arrangements of branch-cuts	71
§ 15.	Every rational function of w and z is like-branched with w .	80

SECTION IV.

INTEGRALS WITH COMPLEX VARIABLES.

§ 16.	Definition of an integral. Dependence of the same upon the path of integration	80
§ 17.	The surface integral $\iint\left(\dfrac{\delta Q}{\delta x}-\dfrac{\delta P}{\delta y}\right)dxdy$ is equal to the linear integral $\int(Pdx+Qdy)$ extended over the boundary. Positive boundary-direction	84
§ 18.	If $Pdx+Qdy$ be a complete differential, then $\int(Pdx+Qdy)$, taken along the boundary of a surface in which P and Q are finite and continuous, is zero. Also, $\int f(z)dz=0$, if this integral be taken along the boundary of a surface in which $f(z)$ is finite and continuous. Importance of simply connected surfaces	90
§ 19.	The value of a boundary integral does not change when pieces in which $f(z)$ is finite and continuous are added to or sub-	

CONTENTS. ix

PAGE

tracted from the bounded surface. A boundary integral is equal to the sum of the integrals taken along small closed lines which enclose singly all the points of discontinuity contained in the surface 93

§ 20. If $f(z)$ become infinite at the point $z = a$ in such a way that $\lim[(z-a)f(z)]_{z=a}=p$, then $\int f(z)dz = 2\pi i p$, integrated round a. Reduction of the values of the integral to closed lines round the points of discontinuity 96

§ 21. Integrals round a branch-point b. If we let $(z-b)^{\frac{1}{m}} = \zeta$ and $f(z) = \phi(\zeta)$, then $\phi(\zeta)$ does not have a branchpoint at the place $\zeta = 0$. If at least $\lim[(z-b)^{\frac{m-1}{m}} fz]_{z=b}$ be finite, then $\int f(z)dz = 0$ 101

SECTION V.

THE LOGARITHMIC AND EXPONENTIAL FUNCTIONS.

§ 22. Definition and properties of the logarithmic function. Multiplicity of values of the same 104

§ 23. The exponential function $z = e^w$. Representation of the z-surface on the w-surface 109

SECTION VI.

GENERAL PROPERTIES OF FUNCTIONS.

§ 24. If $\phi(z)$ be uniform and continuous in a surface T, then for every point t in that surface $\phi(t) = \frac{1}{2\pi i}\int\frac{\phi(z)dz}{z-t}$, the integral to be taken round the boundary of T. In a region in which a function $\phi(z)$ is uniform and continuous, its derivatives are also uniform and continuous, and if we put $\phi(z) = u + iv$, then neither mod u nor mod v can have a maximum or minimum value at any place in this region 113

§ 25. The domain of a point a. A function which is finite and continuous at a point a can be represented by a convergent series of ascending powers of $t - a$ for all points t in the domain of a. Taylor's series. A function of a complex variable can be continued in only one way; if it be constant in any arbitrary small finite region, it is constant everywhere 117

x   CONTENTS.

§ 26. Representation of a function by a convergent series in the domain of a point which is not a branch-point, but at which the function is discontinuous 121

SECTION VII.

INFINITE AND INFINITESIMAL VALUES OF FUNCTIONS.

A. *Functions without branch-points.*   *Uniform functions.*

§ 27. Polar and non-polar discontinuities. A uniform function at a point of discontinuity of the second kind must become infinite and be capable of assuming every arbitrary value. Example 125

§ 28. A uniform function which does not become infinite for any finite or infinite value of the variable is a constant. A uniform function which is not a mere constant must become infinite and zero, and be capable of acquiring any given value 131

§ 29. An infinite value of a definite order. Further characterization of the two kinds of discontinuity 133

§ 30. An infinite value for $z = \infty$ 141

§ 31. A uniform function which becomes infinite only for $z = \infty$ and that of a finite order is an integral function. If, for $z = \infty$, it become infinite of an infinitely high order, it can be developed in a series of powers of z converging for all finite values of the variable 143

§ 32. A uniform function which becomes infinite only a finite number of times is a rational function 144

§ 33. $\phi(z)$ is determined, except as to an additive constant, when for each of its points of discontinuity we are given a function which becomes discontinuous just as $\phi(z)$ does, 145

§ 34. An infinitesimal or zero value 146

§ 35. If in a given region $\phi(z)$ become zero of multiplicity n and infinite of multiplicity ν, then $\int d \log \phi(z) = 2\pi i(n - \nu)$, the integral to be taken round the boundary of the region 147

§ 36. A uniform function in the whole infinite extent of the plane is just as often zero as it is infinite 150

§ 37. A uniform function is determined except as to a constant factor, when once we know all finite values for which it becomes infinite and infinitesimal, and also the order of the infinite or infinitesimal value for each 153

B. *Functions with branch-points.   Algebraic functions.*

§ 38. The infinite value of an algebraic function. It becomes infinite of a fractional order at a branch-point 155
§ 39. Behavior of the derived function at a branch-point 160
§ 40. Representation of the surface in the neighborhood of a branch-point 167
§ 41. An n-valued function, which becomes infinite of multiplicity m, is the root of an algebraic equation between w and z, of the nth degree with regard to w, the coefficients of which are integral functions of z at most of the mth degree 169

SECTION VIII.

INTEGRALS.

A. *Integrals taken along closed lines.*

§ 42. The integral $\int f(z)dz$, taken round a point of discontinuity about which the z-surface winds m times, and at which $f(z)$ becomes infinite of a finite order, has a value different from zero, when, and only when, the term which becomes infinite of the first order is present in the expression defining the nature of the infinite value of $f(z)$; and this value is equal to $2m\pi i$ times the coefficient of this term 172
§ 43. Closed lines round the point at infinity. The integral taken along such a line depends upon the nature of the function $z^2 f(z)$ 175

B. *Integrals along open lines.   Indefinite integrals.*

§ 44. Behavior of the integral of an algebraic function $\phi(z)$, when the upper limit acquires a value for which $\phi(z)$ becomes infinite. Logarithmic infinity 180
§ 45. Behavior of the integral when the upper limit tends towards infinity. Examples 182

SECTION IX.

SIMPLY AND MULTIPLY CONNECTED SURFACES.

		PAGE
§ 46.	Definition. Criterion for determining whether a closed line forms by itself alone the complete boundary of a region. Examples	184
§ 47.	Cross-cuts	189
§ 48.	Preliminary propositions	192
§ 49.	Riemann's fundamental proposition	196
§ 50.	Digression on line-systems	205
§ 51.	Lippich's proposition. Another proof of the fundamental proposition	209
§ 52.	Classification of surfaces according to the order of their connection	212
§ 53.	Various propositions	215
§ 54.	Determination of the order of connection for a closed Riemann surface which possesses no boundary-lines, but only a boundary-point	221
§ 55.	The same for a Riemann surface which is extended over a finite part of the plane	224
§ 56.	The same for an arbitrary Riemann surface which possesses boundary-lines	232
§ 57.	Extension of the Eulerian relation between the number of corners, edges, and faces of a body bounded by plane surfaces when it has an arbitrary form	242

SECTION X.

MODULI OF PERIODICITY.

§ 58.	Examination of a function defined by an integral in a multiply connected surface. On crossing a cross-cut the function changes abruptly by a quantity which is constant along the cross-cut. Multiformity of the functions defined by integrals. The inverse functions are periodic	246
§ 59.	Extension for the case in which previous cross-cuts are divided into segments by subsequent cross-cuts. The number of independent moduli of periodicity is equal to the number of cross-cuts	251
§ 60.	More rigorous determination of the points which must be excluded from the surface in the examination of a function defined by an integral, and which must not be excluded	256

CONTENTS. xiii

§ 61. Examples   PAGE
- 1. The logarithm 258
- 2. The inverse tangent 259
- 3. The inverse sine 260
- 4. The elliptic integral 27

Supplementary note to Riemann's fundamental proposition on multiply connected surfaces 286

ELEMENTS

OF THE

THEORY OF FUNCTIONS OF A COMPLEX VARIABLE.

INTRODUCTION.

To follow the gradual development of the theory of imaginary quantities is especially interesting, for the reason that we can clearly perceive with what difficulties is attended the introduction of ideas, either not at all known before, or at least not sufficiently current. The times at which negative, fractional and irrational quantities were introduced into mathematics are so far removed from us, that we can form no adequate conception of the difficulties which the introduction of those quantities may have encountered. Moreover, the knowledge of the nature of imaginary quantities has helped us to a better understanding of negative, fractional and irrational quantities, a common bond closely uniting them all.

Among the older mathematicians, the view almost universally prevailed that imaginary quantities were impossible. In glancing over the earlier mathematical writings, we meet with the statement again and again that the occurrence of imaginary quantities has no other significance than to prove the impossibility or insolubility of a problem, that these quantities have no meaning, but may sometimes be profitably employed, the form of the results being then merely symbolical. In this connection it is interesting to observe the development of Cauchy's process. This great mathematician, together with the "Princeps mathematicorum," Gauss, who had first, and probably very early, recognized the great importance of imaginary quantities in all parts of mathematics, may be considered the joint-creator of the theory of functions of imaginary variables. Yet, both in his *Algebraical Analysis* and also in the *Exercises* of the year 1844, he still followed entirely the views of the older mathematicians. In one place we read[1]: "Toute équation imaginaire n'est autre chose que la réprésentation symbolique de deux équations entre quantités réelles. L'emploi des expressions imaginaires, en permettant de remplacer deux équations par une seule, offre souvent le moyen de semplifier les calculs et d'écrire sous une forme abrégée des résultats fort compliqués. Tel est même le motif principal pour lequel on doit continuer à se servir de ces expressions, qui prises à la lettre et interprétées d'après les conventions généralement établies, ne signifient rien et n'ont pas de sens. Le signe $\sqrt{-1}$ n'est en quelque sorte qu'un outil, un instrument de calcul, qui peut-être employé avec succès dans un grand nombre de cas pour rendre beaucoup plus simples non-seulement les formules analytiques, mais encore les méthodes à l'aide desquelles on parvient à les établir."

These words indicate very clearly the standpoint of the older mathematicians, which, as may be seen, was still maintained by some at a much later period. In one only of the mathematical branches have imaginary quantities always been recognized, namely, in the theory of algebraical equations; for here it was far too important to consider all the roots together, for the imaginary state of any of the latter to interrupt the investigations. Nevertheless, individual men, as de Moivre, Bernoulli, the two Fagnano, d'Alembert and Euler, who seemed to turn to imaginary quantities with especial

---

[1] Cauchy, *Exercises d'analyse et de physique mathématique*, Tome III. p. 361.

predilection, gradually discovered the distinguishing properties inherent in these quantities, and more and more developed their theory. Still, as a whole, these investigations were looked upon rather as scientific pastimes, as mere curiosities, and were held to be of value only in so far as they lent themselves as aids to other investigations. And there have not been wanting those who opposed the employment of imaginary quantities altogether, on account of their supposed impossibility.[1]

The opinion that imaginary quantities are impossible has its true origin in mistaken ideas of the nature of negative, fractional and irrational quantities. For the application of these mathematical ideas to geometry, mechanics, physics, and partially even to civic life, presenting itself so readily and so spontaneously, and in many cases no doubt even giving rise to some investigation of these quantities, it came to be thought that in some one of these applications should be found the true nature of such ideas and their true position in the field of mathematics. Now, in the case of imaginary quantities, such an application did not readily present itself, and owing to insufficient knowledge of the same it was thought that they should be relegated to the realm of impossibility and their existence be doubted.

But thereby it was overlooked that pure mathematics, the science of addition, however important may be its applications, has in itself nothing to do with the latter; that its ideas, once introduced by complete and consistent definitions, have their existence based upon these definitions, and that its principles are equally true, whether or not they admit of any applications. Whether and when this or that principle will find an application cannot always be determined in advance, and the

---

"[1] Aussi a-t-on vu quelques géomètres d'un rang distingué ne point goûter, ce genre de calcul, non qu'ils doutassent de la justesse de son résultat, mais parce qu'il paraissait y avoir une sorte d'inconvenance à employer des expressions de ce genre qui n'ont jamais servi qu'à annoncer une absurdité dans l'énoncé d'un problême." — MONTUCLA, *Histoire des Mathématiques*, Tome III. p. 283.

present time especially is rich enough in instances in which the most important applications — even those of far-reaching influence on the life of nations — have sprung from principles, at the discovery of which there was certainly no suggestion of such results. But so firm had the belief in the impossibility of imaginary quantities gradually become that, when the idea of representing them geometrically[1] first arose in the middle of the last century, from the supposed impossibility of the same, was inferred conversely the impossibility of representing them geometrically.[2]

To understand the position which imaginary quantities occupy in the field of pure mathematics, and to recognize that they are to be put upon precisely the same footing as negative, fractional and irrational quantities, we must go back somewhat in our considerations.

The first mathematical ideas proceeding immediately from the fundamental operation of mathematics, *i.e.*, addition, are those which, according to the present way of speaking, are called positive integers.

If from addition we next pass to its opposite, subtraction, it soon becomes necessary to introduce new mathematical concepts. For, as soon as the problem arises to subtract a greater number from a less, it can no longer be solved by means of positive integers. From the standpoint in which we deal with only positive integers, we have therefore the alternative, either to declare such a problem impossible, insoluble, and thus to

---

[1] On the history concerning the geometrical representation of imaginary quantities, compare Hankel, *Theorie der complexen Zahlensysteme*, Leipzig, 1867, S. 81. It deserves to be noted that Abel and Jacobi, in opposition to the view that only a geometrical representation could secure for imaginary quantities a real existence, already made unlimited use of imaginary quantities in their first investigations on elliptic functions, and this at a time when that representation was all but unknown. Fully conscious of how essential the consideration of imaginary quantities was, and how incomplete their investigations would remain without them, they disregarded entirely the question of their possibility or impossibility.

[2] Foncenex, "Reflexions sur les quantités imaginaires," *Miscellanea Taurinensia*, Tomo I. p. 122.

put a stop to all further progress of the science in this direction; or, on the other hand, to render the solution of the problem possible by introducing as new concepts such mathematical ideas as enable us to solve the problem. In this way negative quantities at first arise through subtraction as the differences of positive integers, of which the subtrahends are greater than the minuends. Their existence and meaning for pure mathematics, then, is not based upon the opposition between right and left, forward and backward, affirmation and negation, debit and credit, or upon any other of their various applications, but solely upon the definitions by which they were introduced.

Now, although the idea of impossibility is not at all contained in our conceptions of negative quantities, it may happen that the occurrence of negative quantities indicates the impossibility or insolubility of a problem, namely, when the nature of the problem necessarily requires positive quantities for its solution. If, for instance, the following problem be given: Six balls are to be distributed in two urns, so that one shall contain eight more than the other; then the following purely mathematical problem is contained in it: to find two numbers of which the sum is equal to six and the difference to eight. Now, if it merely be desired that the numbers shall be mathematical concepts without limiting them to a special kind, and if, moreover, the conception of negative quantities has been fixed beforehand by defining them, the solubility of the purely mathematical problem is quite obvious — the positive number 7 and the negative number -1 are the quantities which satisfy the problem. Nevertheless, it is impossible to solve the problem originally set, for it requires that each of the numbers sought shall stand for a quantity, and therefore necessarily be positive. If the impossibility were not so obvious as it is in this simple example, the occurrence of the negative number -1 would show conclusively the insolubility of the problem.

Exactly the same conditions arise in every other inverse operation. The next inverse operation is division. If we set the problem to divide a whole number by another which is not a factor of the first, there arises the impossibility of solving this problem by positive or negative integers. The progress of the science therefore again requires the possibility of the solution to be brought about by introducing and defining the quantities necessary to that end. Here these new concepts are rational fractions. But here, too, the case may occur that the appearance of such quantities proves the impossibility of solving a particular problem; and again, as before, when by the nature of the problem it does not admit of a solution in terms of the new concepts. Take as an example the following problem: A wheel in a machine or clock work, which has 100 cogs and revolves once a minute, is to set directly in motion another wheel, so that the latter shall make 12 revolutions in a minute; how many cogs must we give to the second wheel? In this case the underlying purely mathematical problem consists merely in dividing 100 by 12; and if the definition of fractions has once been given, the solution presents no difficulty, the result being $8\frac{1}{3}$. But the occurrence of this fraction proves at once the impossibility of solving the problem originally proposed, as the number of cogs on the second wheel to be determined must be an integer.

The third inverse operation is the extraction of roots.

Given $\sqrt[n]{a} = x$,

in which n denotes a positive integer; the problem to find a quantity x satisfying this equation can no longer be solved in terms of whole numbers or rational fractions, as soon as a is not the nth power of such a quantity. In this case therefore the necessity again arises of rendering the problem soluble by the introduction of new concepts. Now, if either a be positive, or in case a is negative, if n be an odd number, the new concepts to be introduced are irrational quantities; but if a be negative, and n at the same time an even number, the new concepts to be introduced are imaginary quantities. Now it is no more an impossibility to define these latter than to

INTRODUCTION.   7

define irrational quantities, or, to go back still farther, than to define rational fractions and negative quantities, for in none of these definitions do we meet with any inherent inconsistencies. Should such occur, should properties be put in combination with one another which we can prove to be inconsistent, then, it must be admitted, we should have actually to deal with an impossibility. Gauss[1] adduces as an example of such an impossibility a plane rectangular equilateral triangle. And indeed it can be proved that a plane equilateral triangle cannot at the same time be rectangular. Something impossible would therefore actually be proposed. If now, in fact, the occurrence of negative quantities, or of fractions, indicate sometimes the impossibility of particular problems, it is easily conceivable that such an impossibility can also be proved by means of imaginary quantities, as in the following example: A given straight line two units long is to be divided into two such parts, that the rectangle formed by them shall have the area 4. The purely mathematical content of this problem is to find two numbers of which the sum equals 2 and the product 4. If now it be required merely that these numbers shall be mathematical quantities, without specifying the particular kind, then, the definition of imaginary quantities having once been given, the solution presents no difficulty. It leads to the solution of the quadratic equation,

$$x^2 - 2x + 4 = 0,$$

of which the roots are the imaginary quantities

$$1 + \sqrt{-3} \text{ and } 1 - \sqrt{-3}.$$

But if we attempt to satisfy the conditions of the original problem, that the quantities sought shall represent parts of a straight line and hence be real quantities, it is impossible to solve the problem, because the greatest rectangle formed

[1] "Demonstratio nova theorematis omnem functionem algebraicam rationalem integram unius variabilis in factores reales primi vel secundi gradus resolvi posse." — *Inaug. Diss.* p. 4, Note.

by two parts of the line 2 has the area 1, and therefore none can have the area 4; and this impossibility is indicated in this case by the occurrence of imaginary quantities. Montucla[1] has chosen this very example in support of his view that the meaning and origin of imaginary quantities are to be looked for altogether in the impossibility of a problem, because these quantities occur when a problem is given which contains an impossible or absurd condition. We have already seen that exactly the same can be affirmed of negative quantities and fractions, and the words: "Ainsi toutes les fois que la résolution d'un problème conduit à de semblables expressions et que parmi les différentes valeurs de l'inconnue il n'y en a que de telles, le problème, ou pour mieux dire, ce qu'on demande est impossible," and further on, "Le problème, qui conduirait à une pareille équation, serait impossible ou ne présenterait qu'une demande absurde," can be applied almost literally to the two examples adduced above, in which the impossibility of the problem was indicated by a negative number and by a fraction respectively.

It is evident from the foregoing considerations that imaginary, irrational, rational-fractional and negative quantities, have all a common mode of origin, namely, by means of inverse operations, in which their introduction is rendered necessary by the further progress of the science. They all have their existence based upon their definitions, no one of which includes anything impossible; but it may happen that the occurrence of each of them proves the impossibility of solving a given problem, on account of the peculiar character of the same.

Before we take up the subject proper, some remarks on the calculations by means of imaginary quantities may be permitted. Here, too, we can start from quantities related to them. Every time a new concept is introduced into mathematics, it is in many respects absolutely a matter of choice in what way the operations upon which the former concepts

[1] *Histoire des Mathématiques*, Tome III. p. 27.

INTRODUCTION.

depend shall be transferred to the new. For instance, after the definition of powers with positive integral exponents has been derived from the repeated multiplication of a quantity by itself, the question arises as to what is to be understood by a power with a negative exponent. In itself the answer is absolutely a matter of choice, for there is nothing which compels us to understand by it one thing and no other. But if in this and all similar cases we had proceeded quite arbitrarily, and had not been guided by any definite principle, the structure of mathematics would surely have assumed a strange form, and the survey of it enormous difficulty. Mathematics owes its external consistency and the harmonious agreement of all its parts to the adherence to the principle that every time a newly introduced concept depends upon operations previously employed, the propositions holding for these operations are assumed to be valid still when they are applied to the new concepts. This assumption, arbitrary in itself, it is permissible to make, as long as no inconsistencies result from it.[1] Now when this principle is adhered to, the definitions which have been discussed above are no longer arbitrary, but follow as necessary results of that principle. In the case of powers, for instance, it is proved that when m and n are two positive integers, and we assume that $m > n$, then

$$\frac{a^m}{a^n} = a^{m-n}.$$

Now we arbitrarily assume that this theorem remains true also when $m < n$; that is, when $m - n = p$ is a negative number; and it follows that we have to put

$$a^{-p} = \frac{1}{a^p},$$

by which the meaning of a power with a negative exponent is now definitely determined.

[1] This is the same assumption that was called by Hankel the principle of the permanence of the formal laws. *Theorie der complexen Zahlensysteme*, Leipzig, 1867, S. 11.

No further argument is needed to prove that the above principle is of the greatest importance for mathematics, notwithstanding the fact that its assumption is by no means necessary but arbitrary.

We need only realize how the system of mathematics would be constituted, were that principle not adhered to, in order to see at once what distinctions we should be forced to make at each step, and how cumbersome would become the methods of proof. The generalizations of mathematical principles brought about by the prevalence of this principle to the widest extent explain also another phenomenon in the history of mathematics, namely, that for a long time the views in regard to the meaning of divergent series differed so radically. As it had been the habit to accept all mathematical propositions as holding generally, it required some time for the conviction to prevail that in the development of series the results hold only under certain limiting conditions, and that in general on the introduction of infinity into mathematics, the principle stated above does not admit of as unconditional applications as before.

But in transferring mathematical processes to imaginary quantities, the above principle admits of the fullest application, and it has been conclusively proved that thereby no inconsistencies arise. It is not our purpose here to repeat the proof; it may, however, be mentioned that that principle, although in other respects always followed, yet in the case of imaginary quantities has not always and generally been accepted. As late as Euler's time mathematicians were not yet unanimous in regard to the meaning of the product of two square roots of negative quantities. Euler himself taught, conformably with the above principle and as now generally accepted, that, if a and b denote two positive quantities,

$$\sqrt{-a} \cdot \sqrt{-b} = \sqrt{ab};$$

i.e., that the product of these two imaginary quantities is equal to a real quantity. But this view was not generally accepted, and Emerson, an English mathematician, taught on the contrary that we are forced to assume that

$$\sqrt{-a} \cdot \sqrt{-b} = \sqrt{-ab},$$

because it would be absurd to assume that the product of two impossible quantities should not also be impossible; and Hutton says in his Mathematical Dictionary[1] that in his time the views of mathematicians were about equally divided on this point.

One of the remarkable properties possessed by imaginary quantities, is that all can be reduced to a single one, namely, the $\sqrt{-1}$, for which Gauss has introduced the now generally accepted letter i.[2] By means of it, moreover, we can also reduce every imaginary quantity to the form

$$z = x + iy,$$

in which x and y denote real quantities. A quantity of this form Gauss has called a *complex quantity*,[3] divesting this term of the general meaning in which it had sometimes been used before, and according to which it denoted any quantity composed of heterogeneous parts, and employing the term to designate a special heterogeneous compound, in which a quantity consists of a real and an imaginary part connected by addition.

The complex quantities comprise also the real ones, namely, in the case when the real quantity y has the value zero. If, on the other hand, the other real quantity be equal to zero, and z therefore be of the form

$$z = iy,$$

the complex quantity is called a *pure imaginary*. If, in the quantity $z = x + iy$, either one or both of the real quantities

---

[1] Hutton, *Mathematical Dictionary*, 1790.
[2] The first place in which this notation is employed is found, *Disquisitiones arithmeticae*, Sect. VII. Art. 337.
[3] "Theoria residuorum biquadraticorum," *Comment. societatis Gottingensis*, Vol. VII. (ad. 1828–32), p. 90.

x and y be variable, z is called a *complex variable*. In order that this shall assume the value zero, it is necessary for both the real quantities x and y to vanish simultaneously, because it is not possible for the two heterogeneous quantities, the real x and the imaginary iy, mutually to cancel each other. On the other hand, in order that the complex quantity z shall become infinitely large, it suffices if only one of its two real components x and y become infinitely large. Likewise, another interruption of continuity occurs in z as soon as either one of the real quantities x and y suffers such an interruption. But as long as both x and y vary continuously, z is also called a *continuous variable complex quantity*.

Even the consideration of real variables and their functions is materially facilitated and rendered most intelligible by the geometrical representation of the same. In a much higher degree is this the case with complex variables; we will therefore first examine the methods of graphically representing imaginary quantities.

SECTION I.

THE GEOMETRICAL REPRESENTATION OF IMAGINARY QUANTITIES.

1. In order to form a geometrical picture of a real variable, we conceive, as is well known, a point moving on a straight line. On this, which we may call the x-axis, or also the principal axis, we assume a fixed point o (the origin), and represent the value of a variable quantity x by the distance $\overline{op}$ of a point p on the x-axis from the origin o. At the same time attention is paid to the direction of the distance $\overline{op}$ starting from o, a positive value of x being represented by a distance $\overline{op}$ toward one side (say, toward the right, if the x-axis be supposed to be horizontal), a negative value of x by a distance $\overline{op}$ toward the opposite side (toward the left). When now x

IMAGINARY QUANTITIES.   13

changes its value, the distance $\overline{op}$ also changes, the point p changing its position on the x-axis. We can therefore say, either that every value of x determines the position of a point p on the x-axis, or that it determines the length of a definite straight line in either of two directions exactly opposite to each other.

A complex variable quantity $z = x + iy$ depends upon two real variables x and y, which are entirely independent of each other. Hence for the geometrical representation of a complex quantity a range of one dimension, a straight line, will no longer suffice, but a region of two dimensions, a plane, will be required for that purpose. The manner of variation of a complex quantity can then be represented by assuming that a point p of the plane is determined by a complex value $z = x + iy$ in such a way that its rectangular co-ordinates, in reference to two co-ordinate axes, assumed to be fixed in the plane, have the values of the real quantities x and y. In the first place, this method of representation includes that of real variables, for when once z becomes real, and therefore $y = 0$, the representing point p lies on the x-axis. Next, the co-ordinates of the point p can vary independently of each other, just as the variables x and y do, so that the point p can change its position in the plane in all directions. Further, one of the two quantities, x and y, can remain constant, while only the other changes its value, in which case the point p will describe a line parallel to the x- or y-axis. Finally and conversely, for every point in the plane the corresponding value of z is fully determined, since by the position of the point p its two rectangular co-ordinates are given, and therefore also the values of x and y.

Instead of determining the position of the point p representing the quantity z by rectangular co-ordinates x and y, we can accomplish the same by means of polar co-ordinates. For, by putting

$$x = r \cos \phi \text{ and } y = r \sin \phi,$$

we obtain $\qquad z = r(\cos \phi + i \sin \phi).$

14   THEORY OF FUNCTIONS.

The real quantity r, which is always to be taken positively, and which is called the *modulus* of the complex variable z, represents then the *absolute length* of the distance $\overline{op}$ (Fig. 1), and ϕ, called the *amplitude* or *argument* of z, the inclination of that *stroke* to the principal axis. Hence we can also say that a complex quantity $r(\cos\phi + i\sin\phi)$ represents a straight line in length and direction, namely, a straight line of which the length is equal to r, and which forms an angle ϕ with the principal axis. The quantity

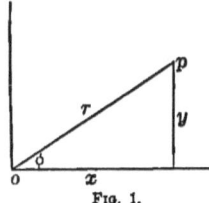

Fig. 1.

$$\cos\phi + i\sin\phi,$$

which depends upon this angle and therefore only upon the direction of the stroke, is usually called the *direction-coefficient* of the complex quantity z.

Just as we can express by a real number any limited straight line, without regarding its direction and position in the plane, or, at most, taking into account only directions exactly opposite to each other; so we can express by a complex quantity a straight line which is determined both in length and direction, but of which the position in the plane is not important. Two given limited straight lines in a plane can actually differ completely in three particulars: in length, direction and position, *i.e.*, the position of that point at which the line is assumed to begin. We can, however, leave out of consideration two of these distinguishing marks, and consider two distances as equal, if they have only equal lengths; this is the case in the representation of distances by real quantities. But in the representation by complex quantities, we dispense with only the third distinguishing mark, namely, the position, and call two distances equal when, and only when, they have equal lengths and directions.

Since the modulus of a complex quantity determines the absolute length of the straight line representing that quantity, it is analogous to the absolute value of a negative quantity

IMAGINARY QUANTITIES.   15

and serves as a measure in comparing complex quantities with one another.

2. From the property of complex quantities that a combination of two or more of them by means of mathematical operations always leads again to a complex quantity, it follows that, if given complex quantities be represented by points, the result of their combination is capable of being again represented by a point. We will now in the following examine the first four algebraical operations,—addition, subtraction, multiplication and division,—and inquire how the points resulting from these operations can be found geometrically. In this the complex quantities, and the points representing them, will always be designated by the same letters; the origin, which represents the value zero, will be designated by o.

1. Addition.

Let $\quad u = x + iy$ and $v = x' + iy'$

be two complex quantities, and let w denote their sum; then

$$w = u + v = (x + x') + i(y + y').$$

The point w therefore has the co-ordinates $x + x'$ and $y + y'$. It follows that it is the fourth vertex of the parallelogram formed on the sides $\overline{ou}$ and $\overline{ov}$, or that by the quantity $u + v$ is represented the diagonal $\overline{ow}$ of this parallelogram in magnitude and direction (Fig. 2). Since the straight lines $\overline{uw}$ and $\overline{ov}$ are equal and directly parallel, and since therefore $\overline{uw}$ is like-

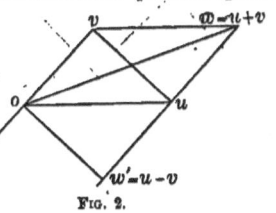

Fig. 2.

wise represented by the complex quantity v, we arrive at the identical point w, if we draw from the end-point u of the first line $\overline{ou}$ the second line $\overline{ov}$ in its given length and direction. This method of combination, or geometrical addition of straight lines, has been applied by Möbius[1] independently of the consideration of imaginary quantities. Accordingly, the sum $u+v$ is the third side of a triangle, of which the two other sides are represented by u and v. Since, however, in every triangle one side is less than the sum of the two other sides, and the lengths of the sides are given by the moduli of the complex quantities, the proposition follows: the modulus of the sum of two complex quantities is less than (or equal to[2]) the sum of their moduli:

$$\mod(u+v) \leqq \mod u + \mod v.{}^3$$

The complex quantity $z = x + iy$ itself appears under the form of a sum of the real quantity x and the pure imaginary iy; since the former is represented by a point on the x-axis, the latter by a point on the y-axis, z is in fact the fourth vertex of the rectangle, the sides of which are formed by the abscissa x and the ordinate y of the point z.

2. Subtraction.

The subtraction of the numbers represented by two points can easily be deduced from the addition of the same; for,

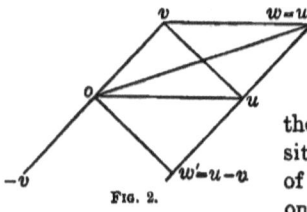

Fig. 2.

given $w' = u - v$,

it follows that

$$u = v + w';$$

therefore the point w' must be so situated that $\overline{ou}$ forms the diagonal of the parallelogram constructed on $\overline{ov}$ and $\overline{ow'}$ (Fig. 2). Consequently, we obtain w' by drawing $\overline{ow'}$ equal and directly parallel to the straight line $\overline{vu}$. Since, however, we pay no attention to the position of a straight line, but only to its

---

[1] Möbius, "Über die Zusammensetzung gerader Linien," etc. *Crelle's Journ.*, Bd. 28, S. 1.
[2] When the moduli of u and v are drawn in the same direction,
  $\mod(u+v) = \mod u + \mod v$. (Translators.)
[3] Mod z (modulus of z) is sometimes denoted by $|z|$. (Tr.)

IMAGINARY QUANTITIES. 17

length and direction, the difference $u - v$ is represented by the straight line $\overline{vu}$ in length and direction (namely, from v to u). The construction shows that u falls in the middle of the straight line $\overline{ww'}$. But from

$$w = u + v, \quad w' = u - v,$$

it follows that

$$u = \frac{w + w'}{2};$$

therefore the point $\frac{w + w'}{2}$ forms the mid-point of the line joining the points w and w'.

If the point u coincide with the origin, *i.e.*, if $u = 0$, then $w' = -v$. In this case a line is to be drawn from o equal in length and direction to the line $\overline{vo}$; hence the point $-v$ lies diametrically opposite to the point v, and equally distant from the origin.

Subtraction furnishes a means of referring points to another origin. For it is evident that a point z is situated with refer-

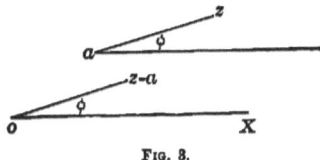

FIG. 3.

ence to a point a exactly as $z - a$ is situated with reference to the origin (Fig. 3).

If we put then

$$z - a = r(\cos\phi + i\sin\phi),$$

r denotes the distance $\overline{az}$, and ϕ the inclination of the line az to the principal axis. The introduction of $z' = z - a$, or the substitution of $z + a$ for z, transfers therefore the origin to a without, however, changing the direction of the principal axis.

3. Multiplication.

We employ here the expression of complex quantities in terms of polar co-ordinates. Let

$$u = r(\cos \phi + i \sin \phi) \quad \text{and} \quad v = r'(\cos \phi' + i \sin \phi')$$

be represented by two points by means of polar co-ordinates, and let w be their product; then

$$w = u \cdot v = rr'[\cos(\phi + \phi') + i \sin(\phi + \phi')].$$

Consequently, the radius vector of w forms with the principal axis the angle $\phi + \phi'$, and its length is equal to the product of the numbers r and r', which denote the lengths of the radii vectors of u and v. From this it follows that the position of the point w, or $u \cdot v$, depends essentially upon the straight line chosen as the unit of length, while the positions of $u + v$ and $u - v$ are independent of this unit. This is quite in accordance with the nature of things, for if in u and v the unit of length be increased in the ratio of 1 to ρ, ρ denoting a real number, the radii vectors of $u + v$ and $u - v$ are increased in the same ratio; the radius vector of $u \cdot v$, however, is increased in the ratio of 1 to ρ^2. Let us assume then on the positive side of the principal axis a point 1, so situated that $\overline{o1}$ is equal to the assumed unit of length (Fig. 4). Since then from the equation

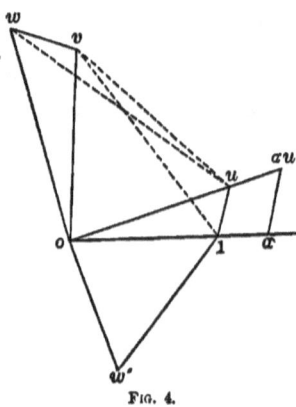

Fig. 4.

$$\overline{ow} = r \cdot r'$$

we obtain the proportion $\quad 1 : r = r' : \overline{ow},$

or $\quad\quad\quad\quad\quad \overline{o1} : \overline{ou} = \overline{ov} : \overline{ow},$

and in addition $\quad\quad \angle vow = \angle 1\,ou,$

IMAGINARY QUANTITIES.

the position of the point w is to be constructed by making the triangles vow and $1\,ou$ directly similar. Instead of this, we could, of course, also make the triangles uow and $1\,ov$ directly similar, which analytically is manifested by the fact that in the product $u \cdot v$ the factors are commutative. From the equation

$$w = u \cdot v$$

we can deduce another proportion, namely,

$$1 : u = v : w;$$

hence the straight lines $\overline{o\,1}$, $\overline{ou}$, $\overline{ov}$, $\overline{ow}$ are proportional to one another, even if their directions be considered. In connection with the preceding, however, it follows that, when straight lines are compared with one another, not only with regard to length but also with regard to direction, two pairs of such lines are proportional, when, and only when, they not only are proportional in length but also in pairs include equal angles; or, in other words, when they are the corresponding sides of directly similar triangles. Now, if we take this requirement into consideration, the last of the above stated propositions serves to find in the simplest manner which triangles have to be made similar to each other; for, from the proportion $1:u = v:w$, it follows by the insertion of the point o that the triangles $1\,ou$ and vow must be similar.

If in the product $u \cdot v$ one of the two factors, say v, be real, and if in this case we denote it by a, then the point representing a lies on the principal axis; hence it follows from the above stated construction that the point representing $a \cdot u$ lies on the line $\overline{ou}$ and at such a distance from o that its radius vector is a times the radius vector of u.

Consequently, the geometrical meaning of multiplication is the following: if a quantity u be multiplied by a real quantity a, the radius vector of u is merely increased in the ratio of 1 to a; but if u be multiplied by a complex quantity v, the radius vector of u is not only increased in the ratio of 1 to mod v, but it is also turned through the angle of inclination of v in the direction in which the arguments increase.

4. Division.

This operation follows immediately from the preceding results. For if

$$w' = \frac{u}{v},$$

we obtain therefrom the proportion

$$w' : 1 = u : v;$$

and hence we have to make the triangles $w'o1$ and uov directly similar (Fig. 4). The geometrical performance of the division of u by v therefore consists in changing the radius vector of u in the ratio of mod v to 1 and, at the same time, in turning it through the angle of inclination of v in the direction in which the arguments decrease.

We will now apply the foregoing considerations to two problems which will be of use to us later.

First: Let z, z' and a be three given quantities, therefore also three given points; we are so to determine a fourth point w that

$$w = \frac{z' - z}{z - a} \quad \text{(Fig. 5)}.$$

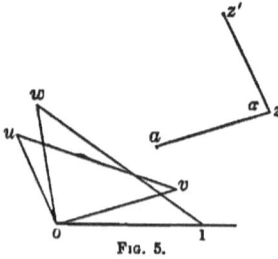

Fig. 5.

If we put $z' - z = u$ and $z - a = v$, we first find the points representing u and v by drawing $\overline{ou}$ equal and parallel to $\overline{zz'}$, and $\overline{ov}$ equal and parallel to $\overline{az}$. We then have $w = \frac{u}{v}$, or $w : 1 = u : v$;

hence we obtain w by making the $\triangle wo1$ similar to $\triangle uov$. From this we can also now deduce for the quantity w an expression which is derived from the sides $\overline{zz'}$ and $\overline{az}$ and the angle azz' of the triangle azz'. For, if this angle be denoted by α, then

$$\angle 1\,ow = \angle vou = 180° - \alpha,$$

moreover,

$$\overline{ow} = \frac{\overline{ou}}{\overline{ov}} = \frac{\overline{zz'}}{\overline{az}};$$

IMAGINARY QUANTITIES. 21

therefore the modulus of w is equal to $\dfrac{\overline{zz'}}{\overline{az}}$, and the direction-coefficient to $(-\cos\alpha + i\sin\alpha)$, and we have

$$w = \frac{\overline{zz'}}{\overline{az}}(-\cos\alpha + i\sin\alpha).$$

In the special case when $\overline{az}$ is perpendicular to $\overline{zz'}$, $\alpha = 90°$, and we obtain

$$w = i\frac{\overline{zz'}}{\overline{az}}.$$

Second: In what relation stand two groups, of three points each, z, z', z'' and w, w', w'', if between them the equation

$$\frac{z'-z}{z''-z} = \frac{w'-w}{w''-w}$$

hold (Fig. 6)? We have immediately the proportion

$$z'-z : z''-z = w'-w : w''-w,$$

and since the differences denote the differences of corresponding points in length and direction, it follows directly that the triangles $z'zz''$ and $w'ww''$ are directly similar.

We here interrupt these considerations, passing over the construction of powers, as not necessary for our purposes. It may, however, be noted that in the case of real integral exponents the construction follows directly from the repeated application of multiplication.[1] One other remark may not be out of place here. If we have an analytical relation between any quantities and carry out the analytical operations on both

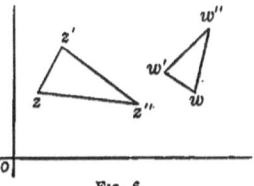

Fig. 6.

---

[1] For powers of any kind we refer to the article: "Ueber die geometrische Darstellung der Werthe einer Potenz mit complexer Basis und complexem Exponenten." (Schlömilch's *Zeitschrift für Mathematik und Physik*, Bd. V., S. 345.)

sides of the equation geometrically, we arrive at the same point by two different methods of construction. Hence, every analytical equation contains at the same time also a geometrical proposition. Thus, for example, it may readily be seen that the identity

$$\frac{u+v}{2} = v + \frac{u-v}{2}$$

furnishes the proposition that the diagonals of a parallelogram bisect each other.[1] By means of a geometrical construction we can also, among other things, render the difference between a convergent and a divergent series quite evident. As is well known, the geometrical progression

$$1 + z + z^2 + z^3 + \cdots$$

has for its sum the value $\dfrac{1}{1-z}$, only when mod $z < 1$. If we now assume an arbitrary point z and construct in the manner given above the points $1, 1+z, 1+z+z^2, 1+z+z^2+z^3$, etc., and if we join these points successively by straight lines, we obtain a broken spiral. If then the point z be so situated that mod $z < 1$, i.e., $\overline{oz} < \overline{o1}$, the points of the spiral approach, on windings which become more and more contracted, that point which can also be obtained by the construction of $\dfrac{1}{1-z}$. But if mod $z \geqq 1$, the windings of the spiral become steadily wider, and an approximation to a fixed point does not occur.

3. The manner of representing geometrically complex values by points in a plane already discussed, also gives us a clear picture of a complex continuous variable. For, if we imagine a series of continuous, successive values of $z = x + iy$, and therefore also a series of continuous successive values of x and y (paired), and if we represent each value of z by a point, these

---

[1] We refer those who wish to follow out still further the line of thought connected with this to the remarkable article by Siebeck: "Ueber die graphische Darstellung imaginärer Funktionen." (*Crelle's Journ.*, Bd. 55, p. 221.)

IMAGINARY QUANTITIES.

points will likewise form a continuous succession, *i.e.*, in their totality a line. Hence, if the variable z change continuously, the point representing z describes a continuous line. Since in this process the real variables x and y can each vary quite independently of the other, the point representing z can also describe an arbitrary line. It deserves to be especially mentioned here that for the continuity of the variation of z it is not at all necessary for the line described by the corresponding point to be a curve proceeding according to one and the same mathematical law, *i.e.*, for the quite arbitrary relation in which x and y must stand to each other in every position of the point to be always expressible by the same equation (or, indeed, by any equation whatever). In order that the variation of z may be continuous, it is necessary only for the line to form a continuous trace. A few examples may make this clear. Suppose the variable z begins its variation with the value $z = 0$, and, after passing through a series of values, acquires a real positive

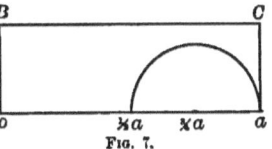

Fig. 7.

value a, which may be represented by the point a (Fig. 7) on the x-axis, the distance $\overline{oa}$ being equal to a. Now the variable z (to express ourselves more briefly, instead of saying, the movable point which represents the corresponding value of the variable z) can pass from o to a on very different paths. *Firstly*, it may assume between o and a only real values, in which case y remains constantly $= 0$ and x increases from 0 to a. The variable describes the straight line oa. *Secondly*, let the variable move along the broken line $oBCa$ formed of three sides of a rectangle in which $oB = b$. In this case x is constantly $= 0$ from o to B, and y increases from 0 to b, so that at B, $z = ib$; then let y maintain the acquired value b, and x increase from 0 to a, so that at C, z assumes the value $a + ib$; finally, from C to a, let x remain constantly $= a$ and y decrease from b to 0. *Thirdly*, the variable z may first move on the principal axis from o to $\frac{1}{2}a$, and then run along a semicircle described round the point $\frac{3}{4}a$ as centre with a radius $\frac{1}{4}a$. This example illustrates at the same time the transference of the origin. On account of the circular motion round the point $\tfrac{3}{4}a$, the course of the real variable becomes much simpler, if we put

$$z - \tfrac{3}{4}a = z' = r(\cos\phi + i\sin\phi).$$

The radii vectors are then measured from the point $\tfrac{3}{4}a$. Now at the origin $z=0$; therefore $z'=-\tfrac{3}{4}a$, and consequently $r=\tfrac{3}{4}a$ and $\phi=\pi$. On the way from o to $\tfrac{1}{2}a$, ϕ remains constantly $=\pi$, and r decreases from $\tfrac{3}{4}a$ to $\tfrac{1}{4}a$, so that at the beginning of the circle $z' = -\tfrac{1}{4}a$, and therefore $z=\tfrac{1}{2}a$. Now in describing the circle, r remains constantly $=\tfrac{1}{4}a$ and ϕ decreases from π to 0, so that at a, $z'=+\tfrac{1}{4}a$, and therefore $z=a$. We have here assumed, as we shall always do in the future, that the angle of inclination ϕ of a complex quantity increases from the direction of the positive x-axis toward the positive y-axis, and we shall call this way of moving *the direction of increasing angles*. From these examples it can be seen that a very essential difference exists between a variable quantity, which is allowed to assume only real values and one which may assume also imaginary values. While by means of two definite values of a real variable, the intermediate series of values, which the variable must assume in order to pass from the first to the second, is completely determined, this is by no means the case with a complex variable; indeed, there are infinitely many series of continuous values, which lead from one given value of a complex variable to another definite value. Geometrically expressed, it may be thus stated: a real variable can proceed only by a single path from one point to another, namely, on the intermediate portion of the principal axis. On the contrary, a complex variable, even when the initial and final values are real, can leave the principal axis and pass from the one point to the other on an infinite number of lines or paths. If the initial and final values, one or both, be complex, the same, of course, holds; and the variable can take arbitrary paths in passing from the one point to the other.

SECTION II.

FUNCTIONS OF A COMPLEX VARIABLE IN GENERAL.

4. In passing next to the consideration of functions of a complex variable, we begin with the elementary idea of a function of a variable quantity, by which is understood any expression formed by the mathematical operations to which the variable is subject; but we shall have to amplify this idea later. In former times, the words "function of a quantity" signified merely what is at present called a power. It is only since the time of John Bernoulli that this term has been applied in its extended meaning, signifying not only the raising to a power, but all kinds of mathematical operations, or any combination of the latter. In more recent times, however, it has become necessary to enlarge still further the concept of a function, and to dispense with the necessity of the existence of a mathematical expression for it. For if one variable be expressed in terms of another, so that the former is a function of the latter, the essential feature of the connection between the two appears in the fact that for every value of the one there is a corresponding value (or several corresponding values) of the other. Now it is this correspondence of the values of the function on the one hand, and of the independent variable on the other, which we especially keep in view. It is also this which is made prominent wherever we recognize the dependence of one quantity upon another, without being able to state the law of this dependence in the form of a mathematical expression. To take a familiar example, we know completely the dependence of the expansion of the saturated vapor of water upon its temperature in such a way that, after the observations made and tables constructed from them, we can determine, within certain limits, the expansion of the vapor for every value of its temperature. But we do not possess a formula derived from theory, by means of which we could calculate the expansion for a given temperature. Notwithstanding, however, the lack of such a mathematical expression, we are still justified

in considering the expansion as a function of the temperature, because to each value of the latter a definite value of the former appertains. The case is the same with algebraic functions in the general sense, *i.e.*, with functions which arise by connecting one variable with another by means of an algebraic equation. As is well known, equations of higher degrees cannot in general be solved, and therefore one variable cannot be expressed in terms of the other. But since we know that to every value of the latter corresponds a definite number of values of the former, we may consider the former as a function of the latter. Besides, functions, whether they admit of being expressed mathematically or not, possess some characteristic properties, usually very small in number, by which they can be determined completely, or at least except as to a constant factor or an additive constant. Hence we can replace the expression of the function by its characteristic properties.

If now we suppose that, within a certain interval of the values of the independent variable, a function is determined only by giving or arbitrarily assuming the value of the latter which corresponds to each value of the former, yet in such a way that, in general, to continuous changes of the variable correspond also continuous variations of the function, then a distinction occurs, according as only real values are assigned to the variable in the given interval, or complex values are also included in the sphere of our discussion. In the former case — the variable assuming only real values — we can, indeed, assume quite arbitrarily the values of the function which are to be attached to those of the variable, and let the one set correspond to the other conformably with continuity. In this case we can always find for the function an analytical expression which shall represent its values within the interval in question; for, if not in any other way, this is always possible by means of the series which proceed according to the sine or cosine of the multiples of an arc. As is well known, this is possible even when the function in isolated places suffers an interruption of its continuity. But when complex values enter into the discussion, we are no longer at liberty to choose arbitrarily a series of continuous complex values, and consider them as the values of a function belonging to a continuous series of values of a complex variable. We shall consider this point more fully later. In the meantime we wish only to call attention to the fact that, even when in a complex variable $w = u + iv$, the quantities u and v are functions of the real constituents x and y of the variable $z = x + iy$, yet w on that account need not be a function of z. We shall first discuss this condition somewhat more fully in the following paragraph.

5. Let us first assume that we have under discussion an expression representing a function of a complex variable $z = x + iy$; then this can be reduced again to the form of a complex quantity, *i.e.*, to the form

$$w = u + iv,$$

wherein u and v denote real functions of x and y. But now every expression of the latter form is not, conversely, at the same time also a function of z; for, that this may be so, it is necessary for the real variables x and y to occur in $u + iv$, only in the definite combination $x + iy$. It is evident that we can easily form functions of x and y in which this is not the case, as, for instance, $x - iy$, $x^2 + y^2$, $2x + iy$. These are, it is true, functions of x and y, but not of $x + iy$; they are *complex functions*, but not *functions of a complex variable*, — concepts, which must therefore be well distinguished. Thus the problem arises to inquire what conditions must be satisfied by a given expression $w = u + iv$, in which u and v signify real functions of x and y, in order that the expression may be a function of $z = x + iy$. To find these conditions, we differentiate w partially as to x and y; then, if w shall in the first place be a function of z, we have

$$\frac{\delta w}{\delta x} = \frac{dw}{dz} \frac{\delta z}{\delta x}$$

$$\frac{\delta w}{\delta y} = \frac{dw}{dz} \frac{\delta z}{\delta y},$$

28   THEORY OF FUNCTIONS.

or, since
$$\frac{\delta z}{\delta x}=1,\ \frac{\delta z}{\delta y}=i,$$

the following
$$\frac{\delta w}{\delta x}=\frac{dw}{dz},\ \frac{\delta w}{\delta y}=i\frac{dw}{dz}.$$

Hence we obtain, as the necessary condition that w shall be a function of z, the equation
$$\frac{\delta w}{\delta y}=i\frac{\delta w}{\delta x}.$$

Conversely, it can easily be proved that this condition is sufficient, *i.e.*, that a function w of x and y, which satisfies this equation, will always be a function of z. For, if in the complete differential
$$dw=\frac{\delta w}{\delta x}dx+\frac{\delta w}{\delta y}dy,$$
we substitute $i\frac{\delta w}{\delta x}$ for $\frac{\delta w}{\delta y}$,

we obtain
$$dw=\frac{\delta w}{\delta x}(dx+idy)=\frac{\delta w}{\delta x}dz.$$

If, however, by means of $z=x+iy$, the variable x be eliminated from the function w before differentiation, and if the partial derivatives as to y and z, derived after the elimination, be distinguished from the former by parentheses, we have
$$dw=\left(\frac{\delta w}{\delta y}\right)dy+\left(\frac{\delta w}{\delta z}\right)dz;$$
by subtracting this expression for dw from the former, we get
$$0=\left\{\frac{\delta w}{\delta x}-\left(\frac{\delta w}{\delta z}\right)\right\}dz-\left(\frac{\delta w}{\delta y}\right)dy.$$

But since dy and dz are entirely independent of each other, separately must
$$\left(\frac{\delta w}{\delta y}\right)=0,\ \left(\frac{\delta w}{\delta z}\right)=\frac{\delta w}{\delta x}.$$

From the first of these equations it follows that w, after the elimination of x, no longer contains y, but is a function of z

only. Then $\left(\dfrac{\delta w}{\delta z}\right)$ and $\dfrac{\delta w}{\delta z}$ have the same meaning, and therefore the second equation gives $\dfrac{dw}{dz} = \dfrac{\delta w}{\delta x}$, the same result as before. Therefore the above relation

(1) $$\dfrac{\delta w}{\delta y} = i\dfrac{\delta w}{\delta x}$$

is the necessary and sufficient condition to ensure the stated functionality of w. From this also follow the equations of condition for the real parts u and v. If $u + iv$ be substituted for w, we obtain

$$\dfrac{\delta u}{\delta y} + i\dfrac{\delta v}{\delta y} = i\left(\dfrac{\delta u}{\delta x} + i\dfrac{\delta v}{\delta x}\right),$$

and then by equating real and imaginary parts,

(2) $$\dfrac{\delta u}{\delta x} = \dfrac{\delta v}{\delta y}, \quad \dfrac{\delta u}{\delta y} = -\dfrac{\delta v}{\delta x}.$$

Finally, we can establish for each of these functions a single equation of condition. For, differentiating each of the above equations partially as to x and y, and eliminating v and u in turn, we obtain

(3) $$\dfrac{\delta^2 u}{\delta x^2} + \dfrac{\delta^2 u}{\delta y^2} = 0 \text{ and } \dfrac{\delta^2 v}{\delta x^2} + \dfrac{\delta^2 v}{\delta y^2} = 0,$$

so that neither of the functions u and v is arbitrary, but each one must satisfy the same partial differential equation. As is well known, partial differential equations do not characterize particular functions but general classes of functions. Thus the function w of the complex variable $z = x + iy$ is given by equation (1), and the real constituent parts of such a function by (2) and (3).

6. If we still hold to the supposition that the function w is given by an expression, an important inference can be drawn from equation (2). To the increment dz of z corresponds the increment $\frac{dw}{dz} dz$ of w. By introducing the quantities u, v and x, y into the derived function $\frac{dw}{dz}$ we obtain

$$\frac{dw}{dz} = \frac{du + idv}{dx + idy} = \frac{\frac{\delta u}{\delta x} dx + \frac{\delta u}{\delta y} dy + i\left(\frac{\delta v}{\delta x} dx + \frac{\delta v}{\delta y} dy\right)}{dx + idy}$$

But now, when the variable z is represented by a point in the xy-plane, this point can move in any arbitrary direction, and the differential
$$dz = dx + idy$$
represents the infinitely small straight line which indicates the change of place of z in magnitude and direction. This infinitely small straight line can therefore be drawn from z in any arbitrary direction. Now, however, the preceding expression for $\frac{dw}{dz}$ shows that it is not independent of dz, but changes its value with the direction of dz. To make it still clearer, let us introduce the differential coefficient $\frac{dy}{dx}$, which indicates the direction of dz, into the expression for $\frac{dw}{dz}$. Dividing numerator and denominator by dx, we obtain

$$\frac{dw}{dz} = \frac{\frac{\delta u}{\delta x} + \frac{\delta u}{\delta y}\frac{dy}{dx} + i\left(\frac{\delta v}{\delta x} + \frac{\delta v}{\delta y}\frac{dy}{dx}\right)}{1 + i\frac{dy}{dx}}; \qquad (4)$$

from which it follows that $\frac{dw}{dz}$, in fact, changes its value with that of $\frac{dy}{dx}$, when no relation exists between the four differential coefficients $\frac{\delta u}{\delta x}$, $\frac{\delta u}{\delta y}$, $\frac{\delta v}{\delta x}$, $\frac{\delta v}{\delta y}$. But if we take the equations (2) into consideration and by means of them eliminate, say, $\frac{\delta u}{\delta y}$ and $\frac{\delta v}{\delta y}$, we obtain

$$\frac{dw}{dz} = \frac{\left(\frac{\delta u}{\delta x} + i\frac{\delta v}{\delta x}\right)\left(1 + i\frac{dy}{dx}\right)}{1 + i\frac{dy}{dx}} = \frac{\delta u}{\delta x} + i\frac{\delta v}{\delta x};$$

FUNCTIONS OF A COMPLEX VARIABLE. 81

thus $\dfrac{dw}{dz}$ becomes independent of $\dfrac{dy}{dx}$ and hence also of dz. *If therefore w be a function of the complex variable $z = x + iy$, the derivative $\dfrac{dw}{dz}$ is independent of dz and has the same value in whatever direction the infinitely small movement may take place.* If we call the different paths which the variable may take the modes of variation, we can say that the derivative is independent of the mode of variation of the variable z. In the case of a function of a real variable, the change of the variable itself does not make any essential difference, because this change can only consist of an increase or decrease of the variable. In the case of functions of a complex variable, however, the different ways in which the variable can change play an important part, and hence the proposition just established, that the derivative of a function of a complex variable is independent of the mode of variation of the variable, is of great importance. And it is only when $\dfrac{dw}{dz}$ is completely independent, *i.e.*, both of the length and of the direction of this infinitely small straight line, that the idea of the derived function becomes as definite as it is in the case of real variables.

Until now we have been assuming that the function w is given by a mathematical expression in terms of z. If we now give up this assumption, we must, in order that the derivative of the function w may have a definite meaning, still add the requirement that it be independent of the differential dz.

The fulfilment of this requirement, however, is sufficient to characterize w as a function of $x + iy$, for from it follow again our former conditions (1), (2) and (3). If the expression (4) for $\dfrac{dw}{dz}$ is to be independent of dz, or what is the same thing, of $\dfrac{dy}{dx}$, the equation resulting from it,

$$\frac{dw}{dz} - \frac{\delta u}{\delta x} - i\frac{\delta v}{\delta x} + \left(i\frac{dw}{dz} - \frac{\delta u}{\delta y} - i\frac{\delta v}{\delta y}\right)\frac{dy}{dx} = 0,$$

must be satisfied for every value of $\frac{dy}{dx}$. Therefore, we obtain

$$\frac{dw}{dz} = \frac{\delta u}{\delta x} + i\frac{\delta v}{\delta x} = \frac{\delta w}{\delta x},$$

$$i\frac{dw}{dz} = \frac{\delta u}{\delta y} + i\frac{\delta v}{\delta y} = \frac{\delta w}{\delta y},$$

or, as above, $\qquad \dfrac{\delta w}{\delta y} = i\dfrac{\delta w}{\delta x}.$

In accordance with this, Riemann[1] has defined a function of a complex quantity in the following way: "*A variable complex quantity w is called a function of another variable complex quantity z, if it so change with the latter that the value of the derivative $\dfrac{dw}{dz}$ is independent of the value of the differential dz.*" Or, as it is expressed in another place[2]: "*If w change with $x + iy$ in conformity to the equation $\dfrac{\delta w}{\delta y} = i\dfrac{\delta w}{\delta x}$*"

It can also be easily proved that, if w be a function of z, the derivative $\dfrac{dw}{dz}$ must likewise be a function of z. For, from the equations

$$\frac{dw}{dz} = \frac{\delta w}{\delta x} = \frac{1}{i}\frac{\delta w}{\delta y}$$

follow $\qquad \dfrac{\delta}{\delta x}\left(\dfrac{dw}{dz}\right) = \dfrac{1}{i}\dfrac{\delta^2 w}{\delta x \delta y}$

and $\qquad \dfrac{\delta}{\delta y}\left(\dfrac{dw}{dz}\right) = \dfrac{\delta^2 w}{\delta x \delta y};$

consequently $\qquad \dfrac{\delta}{\delta y}\left(\dfrac{dw}{dz}\right) = i\dfrac{\delta}{\delta x}\left(\dfrac{dw}{dz}\right),$

and therefore $\dfrac{dw}{dz}$ also satisfies equation (1).

---

[1] *Grundlagen für eine allgemeine Theorie der Funktionen einer veränderlichen complexen Grösse.* S. 2.
[2] "Allgemeine Voraussetzungen," etc., *Crelle's Journ.*, Bd. 54, S. 101.

FUNCTIONS OF A COMPLEX VARIABLE. 33

Further, if w be a function of $z = x + iy$, and if z be a function of $\zeta = \xi + i\eta$, then w is also a function of ζ. For, as above, p. 28,

$$dw = \frac{\delta w}{\delta x}(dx + idy) = \frac{\delta w}{\delta x}dz,$$

and similarly $\quad dz = \dfrac{\delta z}{\delta \xi}(d\xi + id\eta),$

consequently $\quad dw = \dfrac{\delta w}{\delta x}\dfrac{\delta z}{\delta \xi}(d\xi + id\eta);$

thus the partial differential coefficients of w as to ξ and η are

$$\frac{\delta w}{\delta \xi} = \frac{\delta w}{\delta x}\frac{\delta z}{\delta \xi}, \quad \frac{\delta w}{\delta \eta} = i\frac{\delta w}{\delta x}\frac{\delta z}{\delta \xi},$$

hence $\quad \dfrac{\delta w}{\delta \eta} = i\dfrac{\delta w}{\delta \xi};$

and therefore w is also a function of $\xi + i\eta$.

7. The condition just established possesses a definite geometrical meaning, which remains to be discussed.

If, as above,

$$z = x + iy \text{ and } w = u + iv,$$

x and y are the rectangular co-ordinates of a point in the z-plane, and u and v the rectangular co-ordinates of a point w in the same or in another plane. If, now, w be a function of z, the position of the point w depends upon the position of the point z, and if z describe a curve, w describes a curve depending upon the latter; in short, if w be a definite function of z, the entire system consisting of the points w is in a definite dependence upon the system formed by the points z. Riemann calls then the system of the points w the *conformal representation* of the system of the points z. In accordance with the above condition, the two

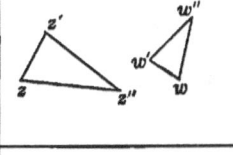

Fig. 6.

figure-systems stand in a quite definite relation to each other, which always holds when w is a function of z.

Let z' and z'' (Fig. 6) be two points infinitely near to a third point z, and let the infinitely small strokes joining them and running in different directions be

$$\overline{zz'} = dz', \quad \overline{zz''} = dz''.$$

Further, let the points which correspond to the points z, z', z'' be w, w', w'', and let the infinitely small strokes joining the latter be

$$\overline{ww'} = dw', \quad \overline{ww''} = dw''{}^{1}.$$

If, now, $\dfrac{dw}{dz}$ is to have the same value for every direction of dz, then

$$\frac{dw'}{dz'} = \frac{dw''}{dz''}, \text{ or } \frac{dw'}{dw''} = \frac{dz'}{dz''}.$$

But the differentials can now be replaced by the differences of the infinitely near points; that is,

$$dz' = z' - z, \quad dw' = w' - w,$$

$$dz'' = z'' - z, \quad dw'' = w'' - w,$$

and we have

$$\frac{w' - w}{w'' - w} = \frac{z' - z}{z'' - z};$$

therefore, by § 2, the triangles $z'zz''$ and $w'ww''$ are similar to each other; *i.e.*, the angles $z'zz''$ and $w'ww''$ are equal to each other, and the included sides are proportional. But since this must hold for any pair of corresponding points z and w, the figure described by the point w is *in its infinitesimal elements similar* to that described by the point z, and two intersecting curves in the w-plane form with each other the same angle as that formed by the corresponding curves in the z-plane. In this connection it must be noticed that $\dfrac{dw}{dz}$ is supposed to be

---

[1] We note that, even if $\dfrac{dw}{dz}$ be independent of dz, yet dw, which $= \dfrac{dw}{dz} dz$, in general changes its direction and magnitude with dz.

neither zero nor infinite. We shall see later that these cases are exceptions.[1] Siebeck terms the dependence of the system w upon the system z *conformation;* and on account of the property that any two pairs of corresponding curves include equal angles, *isogonal conformation.*[2] The simplest isogonal conformations are *similarity* and *circular conformation*[3] (introduced into geometry by Möbius). In the former, $w = az + b$;· in the latter, $w = \dfrac{az+b}{cz+d}$, wherein a, b, c, d denote constants. *Collinearity* and *affinity* are not isogonal conformations; these do not admit of being represented by functional relations between two complex variables.

The simple function $\quad w = z^2$

may serve as an example.

We obtain here $\quad w = x^2 - y^2 + 2\,ixy$,

and hence $\quad u = x^2 - y^2, \quad v = 2\,xy$,

$$\frac{\delta u}{\delta x} = 2\,x, \qquad \frac{\delta v}{\delta x} = 2\,y,$$

$$\frac{\delta u}{\delta y} = -2\,y, \qquad \frac{\delta v}{\delta y} = 2\,x,$$

which verify the equations of condition (2). Let now z describe, for instance, the y-axis, so that $x = 0$, then $z = iy$ and $w = -y^2$; hence w describes the negative part of the principal axis and only this, so that, when z goes from a through o to b, w moves from a' to o, and then back again to b'; a' and b' coincide when $\overline{ao}$ is assumed equal to $\overline{ob}$ (Fig. 8). Let z further describe a circle with radius r, round the origin, so that when

$$z = r(\cos\phi + i\sin\phi),$$

r remains constant; then

$$w = r^2(\cos 2\phi + i\sin 2\phi),$$

---

[1] Cf. § 40.
[2] Known also as *isogonal* or *orthomorphic transformation.* (Tr.)
[3] Called also *bilinear* or *homographic transformation.* (Tr.)

and w also describes a circle round the origin with radius r^2. But since the angle 2ϕ of the w-plane corresponds to the angle ϕ of the z-plane, w describes its circle twice as rapidly as z. For instance, if z describe a semicircle from a to b in the direction of increasing angles, w describes a complete circle from a' to the point b' (which coincides with a'). But the angles, which the straight line and the circle form with

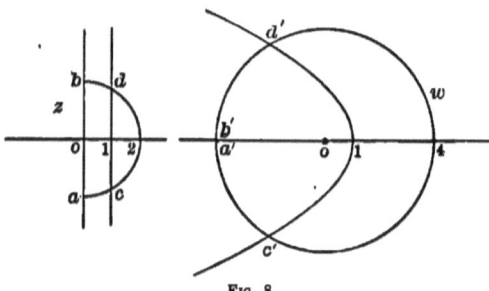

Fig. 8.

each other in z and in w, are both right angles. If we let z describe a straight line cd passing through the point 1 and parallel to the y-axis, w will describe a parabola. This result can be easily obtained,—since in this case x is constant and equal to 1,—by substituting the value of x in the equations $u = x^2 - y^2$ and $v = 2xy$ and eliminating y; thereby we obtain the equation $v^2 = 4(1-u)$ between the co-ordinates u and v of the point w, which shows that the locus of w is a parabola, with its vertex at 1, its focus at o, and of which the parameter, the ordinate at the focus, is 2. By examining the tangents at the points of intersection c' and d', which correspond to c and d, it is easily verified that the parabola cuts the circle in the w-plane under the same angle as the straight line cd cuts the circle in the z-plane. But finally, in order to illustrate an exception by an example, let z describe the principal axis; then z remains real, hence w is positive and therefore describes the positive part of the principal axis. But the latter forms with the negative part of the principal axis, which corresponds to the y-axis in the z-plane, an angle of 180°, while the x- and y-axis in the z-plane form an angle of 90°. Therefore, in the vicinity of the origin the similarity of infinitesimal elements does not occur, and, in fact, at this point the derivative $\dfrac{dw}{dz} = 2\,z$ becomes zero.

SECTION III.

MULTIFORM FUNCTIONS.

8. The introduction of complex variables also throws a clear light on the nature of *multiform* (*many-valued*) functions. For, since a complex variable may describe very different paths in passing from an initial point z_0 to another point z_1, the question naturally suggests itself, whether the path described cannot affect the value w, which a function, starting with a definite value w_0 corresponding to z_0, acquires at the terminal point z_1; we have to inquire whether the curves described by w, starting from w_0, which correspond to those described between z_0 and z_1, must always end in the same point w_1, or whether they cannot also end in different points. Now, in the first place, it is clear that, in the case of *uniform* (*one-valued*) functions, the final value w_1 must be independent of the path taken; for, otherwise, the function would be capable of assuming several values for one and the same value of z, which is not possible with uniform functions. This reason, however, does not apply in the case of multiform functions. Such a function has, in fact, several values for the same value of z, and hence the possibility, that different paths may also lead to different points or to different values of the function, is not excluded at the outset. Let the variable z in $w = \sqrt{z}$, for instance, pass from 1 to 4 by different paths, and let the function w start with $w = +1$ corresponding to $z = 1$; then it is possible that some of the paths shall lead from $w = +1$ to

$w = +2$, and others, on the contrary, from $w = +1$ to $w = -2$. This is, indeed, actually the case in this example. For let $z = r(\cos\phi + i\sin\phi)$, then $w = \sqrt{r}(\cos\tfrac{1}{2}\phi + i\sin\tfrac{1}{2}\phi)$, in which by $\sqrt{r}$, since it is the modulus of w, is to be understood the positive value of the square root of r. Since w is to start with the value $+1$, the initial values of the real variables are $r = 1$ and $\phi = 0$. If z describe a path between 1 and 4, which does not enclose the origin,—for instance, the straight line from 1 to 4,—then ϕ arrives at the point 4 with the value zero, while r acquires here the value 4; hence along such a path w receives the value $+2$. If, on the other hand, the path described by z between 1 and 4 go once round the origin, then at the point 4, ϕ acquires the value 2π, and $\tfrac{1}{2}\phi$ the value π, while again $r = 4$; hence w acquires in this case the value -2. (Cf. also § 10.)[1]

Here we must first of all direct our attention to those points, at which two or more values of the function w, in general different, become equal to one another. Such a point is, for instance, $z = 0$ for $w = \sqrt{z}$; at this point the values of w, in general of different signs, become equal to zero.

Let us next consider the function defined by the cubic equation

$$w^3 - w + z = 0.$$

If, for brevity, we put

$$p = \sqrt[3]{\tfrac{1}{2}(-z - \sqrt{z^2 - \tfrac{4}{27}})}, \quad q = \sqrt[3]{\tfrac{1}{2}(-z + \sqrt{z^2 - \tfrac{4}{27}})},$$

and the two imaginary cube roots of unity

$$\frac{-1 + i\sqrt{3}}{2} = a, \quad \frac{-1 - i\sqrt{3}}{2} = a^2,$$

Cardan's formula gives for the three roots of the above

---

[1] We have in view in these considerations the (irrational) algebraic functions, and hence always assume that the number of values which the function can assume for the same value of the variable z is finite.

equation, which may be denoted by w_1, w_2, w_3, the following expressions:

$$w_1 = p + q,$$
$$w_2 = ap + a^2 q,$$
$$w_3 = a^2 p + aq.$$

For each value of z, w has in general the three values w_1, w_2, w_3. But the last two of these become equal when $p = q$, which occurs when

$$z = \frac{2}{\sqrt{27}}.$$

At this point we have

$$w_2 = w_3 = \sqrt{\tfrac{1}{3}}.$$

If now, in further discussion of this example, we assume that the variable z changes continuously, or that the point representing it describes a line, then each of the three quantities, $w_1, w_2, w_3,$ likewise changes continuously, or the three corresponding points describe three separate paths. But when z passes through the point $z = \frac{2}{\sqrt{27}}$, both functions, w_2 and w_3, assume the value $\sqrt{\tfrac{1}{3}}$; hence the two lines described by w_2 and w_3 meet in the point $\sqrt{\tfrac{1}{3}}$. At the passage through this point therefore w_2 can go over into w_3, and w_3 into w_2, without interruption of continuity; indeed, it remains entirely arbitrary on which of the two lines each of the quantities w_2 and w_3 shall continue its course. In this place a branching, as it were, of the lines described by the quantities w_2 and w_3 takes place; hence Riemann has called those *points of the z-plane*, at which one value of the function can change into another, *branch-points*. In our example therefore $z = \frac{2}{\sqrt{27}}$ is a branch-point (not $w = \sqrt{\tfrac{1}{3}}$). Figures A and B are added in explanation. In Fig. A the three lines w_1, w_2, w_3 are drawn for the case when z describes a straight line parallel to the y-axis and passing through the branch-point $e = \frac{2}{\sqrt{27}}$ (Fig. B). Therein, however,

40   THEORY OF FUNCTIONS.

the line w_1 is represented, for clearness, on twice as large a scale as the remaining lines and, to save space, it is drawn nearer to the ordinate axis than it really runs. The w-points which correspond to the z-points are denoted by the same letters with attached subscripts 1, 2, 3. The picture of the

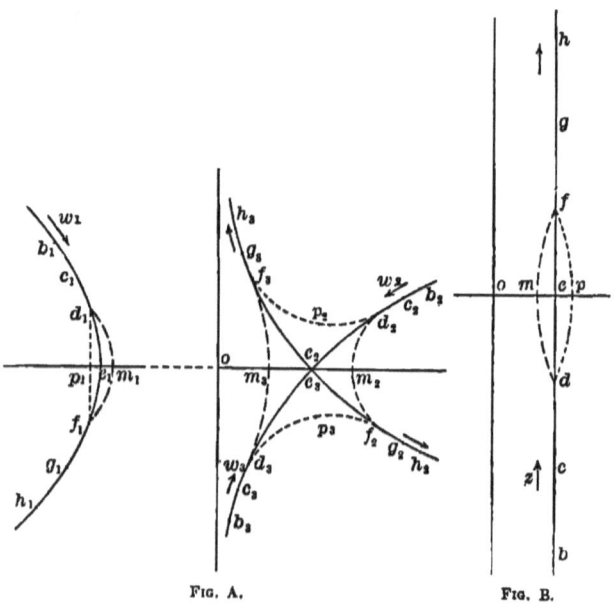

FIG. A.   FIG. B.

branching is rendered still clearer by following the path of only one of the quantities, say w_3. This describes the line $b_3 c_3 d_3$, and approaches the point $e_3 = e_2 = \sqrt{\frac{1}{3}}$, as z approaches the point $e = \dfrac{2}{\sqrt{27}}$ along the line bcd; should z now pass through this point, w_3 could continue its course from

$$e_2 = e_3 = \sqrt{\tfrac{1}{3}}$$

on either of two paths, namely, $e_3f_3g_3h_3$ or $e_2f_2g_2h_2$, of which one as well as the other can be considered the path corresponding to the continuation $efgh$ of z; the way open to w_3 is in fact divided at $e_2 = e_3$ into two branches. When z goes from b to h through the branch-point e, then w_3, starting from b_3, can arrive at h_2 just as well as at h_3; and the same is true of w_2 starting from b_2. In case the path of z leads through the branch-point, the final value of the function remains therefore undetermined. If, on the contrary, z describe a path from b to h, which does not pass through a branch-point, the final value of the function may, it is true, differ according to the nature of the path, but it is for each definite path of z always completely determined. The figures A and B illustrate this also. If z move from b through d, and next along the broken line through m to f and h, then w_3 moves from b_3 through d_3, and next along the broken line through m_3 to f_3 and h_3; w_2 moves from b_2 through d_2, m_2, f_2 to h_2; w_3 acquires then the definite value h_3, and w_2 the definite value h_2. These final values will be different, but again definite, when z goes round the branch-point e on the other side along the dotted line through p. In this case w_3 goes from b_3 through d_3, and then along the dotted line through p_3 to f_2 and h_2; and w_2 goes through d_2, p_2, f_3 to h_3. In this case the successive values of the function, and therefore also the final values, are different from the former, but again they are completely determined.

As a general rule, only those points of the z-plane, at which several values of the function (elsewhere unequal) become equal, are also branch-points. An exception to this is to be mentioned immediately.

A similar branching of the function takes place at those points, at which w becomes infinite and therefore discontinuous. Thus, for instance, the point $z = 0$ is a branch-point both for the function $w = \dfrac{1}{\sqrt{z}}$ and for $w = \sqrt{z}$. Further, in the function determined by the equation

$$(z-b)(w-c)^3 = z-a \text{ or } w = c + \sqrt[3]{\dfrac{z-a}{z-b}},$$

in which a, b, c denote three complex constants, and therefore three points, $z = a$ is a branch-point, at which all three values of the function become equal to $w = c$. Moreover, at $z = b$ all three values of w become infinite. The three functions suffer here an interruption of continuity, and hence it can remain undetermined on which path each is to continue its course, because, when the function makes a spring, it can just as well spring over to the one as to the other continuation of its path. Therefore $z = b$ is likewise a branch-point. Also, as a general rule, those points at which w becomes infinite or discontinuous are branch-points. Exceptions to this, however, may occur; there are cases in which points are not branch-points, although at them the values of the functions are either equal or infinite. This, for the present, can only be illustrated by examples. In the functions

$$\sqrt{1-z^2} \text{ and } \frac{1}{\sqrt{1-z^2}},$$

$z = +1$ and $z = -1$ are branch-points; on the contrary, in

$$(z-a)\sqrt{z} \text{ and } \frac{1}{(z-a)\sqrt{z}},$$

$z = a$ is not a branch-point, although the values of the functions at this place are in the first case both zero, and in the second case both infinite. For, when z passes through the point a, then $z - a$ as well as $\sqrt{z}$ has a perfectly definite, continuous progress; $z - a$, because it is uniform, and $\sqrt{z}$, because $+\sqrt{a}$ cannot, without interruption of continuity, suddenly pass over into $-\sqrt{a}$. Hence the rational functions of these quantities have at this point a definite continuation for every path described by z, and there is no branching. Accordingly, the branch-points are to be looked for only among those points at which an interruption of continuity occurs, or at which several values of the function become equal; but whether such points are actually branch-points must still be expressly determined.

MULTIFORM FUNCTIONS. 43

9. The preceding considerations have shown that, when the variable z starting from an arbitrary point z_0 describes a path to another point z_1, which leads through a branch-point of a function w, the latter acquires different values at z_1 according as it is allowed to proceed on one or another of its branches. Therefore, in the case of such a path of z, the value of w at z_1 is undetermined. If, on the contrary, z describe any other path, not leading through a branch-point, w acquires at z_1 a definite value, and it will now be shown that two paths, both of which lead from z_0 to z_1, *assign different values to w at z_1, only when they enclose a branch-point*. To that end we first prove the following proposition:

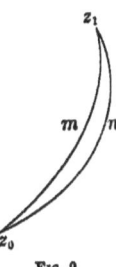

Fig. 9.

Let the variable z in passing from z_0 to z_1 describe two infinitely near paths, z_0mz_1 and z_0nz_1 (Fig. 9), *which in no place approach infinitely near a point at which either the function w becomes discontinuous or several values of the function become equal, then the function w, starting from z_0 with one and the same value, acquires at z_1 the same value on both paths.*

To prove this proposition, we first remark that the different values which a multiform function has at one and the same point z can differ by an infinitely small quantity, only when the point z lies infinitely near a point at which several values of the function become equal. (Cf. Figs. A and B, p. 40. In that example the lines described by the values of the function approach each other only at the point e, while at all other points z they are a finite distance apart.) Since now, according to the hypothesis, the two paths, z_0mz_1 and z_0nz_1, nowhere approach such a point, the different values which w can have at any point of the two paths differ by a finite quantity. Therefore the values which the function w acquires at z_1 on the two paths, z_0mz_1 and z_0nz_1, must either be equal to each other or differ by a finite quantity. But the latter alternative cannot occur. For, if we suppose that two movable z-points describe the two infinitely near paths, z_0mz_1 and z_0nz_1, in such a way that they remain always infinitely near each other, and if we denote the value of the function along the one line by w_m, and that along the other by w_n, then w_m and w_n along both lines can differ only by an infinitely small quantity, because by the hypothesis w starts from z_0 with the same value on both paths, and on both changes continuously, and because, further, in passing from a point of one line to an infinitely near point of the other, the continuity is not broken. Now, if w_m and w_n differed by a finite quantity at z_1, at least one of these functions would have to make a spring in some place, which is excluded by the hypothesis that the two paths, $z_0 m z_1$ and $z_0 n z_1$, shall approach no point at which an interruption of continuity occurs. Consequently w_m and w_n cannot differ from each other by a finite quantity, and hence, according to the above, they are equal to each other.

This having been established, if we now suppose a series of successive paths lying infinitely near to each other, all between the points z_0 and z_1, and so constructed that no one of them approaches a point at which either discontinuity occurs or function-values become equal, then the function acquires on all these paths the same value at z_1. From this follows the proposition: *If a path between two points, z_0 and z_1, can be so deformed into another path by gradual changes, that thereby no one of the above defined critical points is passed over, then the function acquires at z_1 the same value on the second path as on the first.* This conclusion holds also in the case when two points, z_0 and z_1, coincide, and when therefore the variable describes a closed line. The above condition is then changed into this: the closed line is not to include any of the critical points mentioned. Hence, if we let the variable z starting from z_0 describe a closed line and return again to z_0, the function acquires here the same value that it had at the beginning, if the closed line include no point at which either discontinuity occurs or function-values become equal.

Such closed lines described by the variable z are highly important in the investigation of the influence which the path followed by the variable z, on its way to any point, exerts on the value which the function w acquires at that point. If a closed line include none of the points already so often mentioned, the function, as has been shown, does not change its value; but if it enclose such a point, the function may, or may not, change its value. Further, if two paths be described by the variable between two points, which enclose no point of that kind, these lead to the same function-value. Hence we have to consider only paths which enclose such a point. Now let a (Fig. 10) be a point of this kind, and assume two paths bdc and bec, which enclose a but no other similar point. Let w start from b with the value w_0 and acquire at c the value W along the path bdc. Then if we let the variable z, before it enters on the other path bec, describe a closed line $bghb$ round the point a, the path $bghbec$ can be deformed into bdc without passing over the point a; therefore w acquires at c likewise the value W along this path, if it start from b with the value w_0. We have therefore the following:

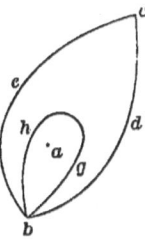

Fig. 10.

along bdc, w changes from w_0 to W,

" $bghbec$, w " " " w_0 " W.

If we first assume that w changes its value by the description of the closed line $bghb$ and goes into w_1, we have:

along $bghb$, w changes from w_0 to w_1,

and hence " bec, w " " " w_1 " W.

Accordingly w acquires at c the value W along bec, when it starts from b with the value w_1; therefore, if it start from b with the value w_0, it cannot acquire the value W, but must be led to another value. If, on the contrary, w do not change its value on the closed line $bghb$, we have:

along $bghb$, w changes from w_0 to w_0,

and hence " bec, w " " " w_0 " W;

therefore in that case w, starting from b with the value w_0, acquires the value W also along the path bec.

From this follows: If two paths enclose one of the points a in question, they lead to different or to the same function-values, according as the function w does or does not change its value in describing a closed line round the point a.

We are now in a position to define branch-points more precisely. A point a, at which either a discontinuity occurs or several function-values become equal, *is to be called a branch-point when, and only when, the function changes its value in describing a closed line round this (and no other similar) point.* Nevertheless, in this connection, it is to be noted that it is not necessary for all the function-values to change. In order that the point in question may be a branch-point, it is only necessary for this change to occur in the case of some one of the function-values under consideration. For the case can occur that, in the circuit round a branch-point, only a part of the function-values change, while the others remain unchanged. The example considered on p. 38 ff. furnishes such a case. Let the variable z, in Fig. B, describe the closed line $dpfmd$, which encloses the branch-point $e = \dfrac{2}{\sqrt{27}}$, then it is evident from Fig. A that w_2 goes over into w_3, and w_3 into w_2; while w_1, however, does not change its value but describes likewise a closed line. Thus the proposition enunciated at the beginning of this paragraph, that two different paths connecting the same point assign different values to a function, which starts from the initial point with the same value, only when they enclose a branch-point, is proved; and for closed lines, we can enunciate the proposition: A multiform function can pass from a value corresponding to a point z_0 to another value corresponding to the same point in a continuous way, when the variable z starting from z_0 describes a closed line which encloses a branch-point.

Closed lines, which enclose two or more branch-points, can likewise be reduced to such closed lines as contain only one branch-point. For, if we draw from a point z_0 a closed line round each branch-point and let the variable describe the same in succession, then this path can be deformed, without passing over one of the branch-points, into a closed line which, starting from z_0, encloses all the branch-points. (Fig. 11, where a and b denote two branch-points.) We draw such closed lines round the individual branch-points most simply, by describing round each one a small circle, and connecting each of these circles with z_0 by a line, which must then be described twice, going and coming.

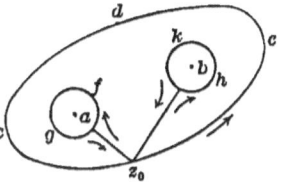

Fig. 11.

10. We will next illustrate the preceding considerations by some examples, and at the same time show by them how the function-values pass into one another on describing closed lines round a branch-point.

Ex. 1. $\qquad w = \sqrt{z}.$

In this $z = 0$ is a branch-point. If we let the variable start from the point $z = 1$ and describe the circumference of a circle round the origin, this is a closed line which encloses the branch-point. If the function $w = \sqrt{z}$ start from the point $z = 1$ with the value $w = +1$, and if we put

$$z = r(\cos \phi + i \sin \phi),$$

then at the point $z = 1$, $r = 1$ and $\phi = 0$. If z next describe the circumference of the circle in the direction of increasing angles, r remains constant and equal to 1, and ϕ increases from 0 to 2π. If therefore the variable return to the point $z = 1$, then

$$z = \cos 2\pi + i \sin 2\pi,$$

and therefore

$$w = \sqrt{z} = \cos \pi + i \sin \pi = -1;$$

thus the function does not now have at the point $z = 1$ the original value $+1$, but acquires the other value -1. The very same takes place, when the variable describes any other closed line once round the origin starting from $z = 1$; for this path can be deformed into the circle by gradual changes without thereby passing over the origin. In general, if w start with the value w_0 from any point z_0, for which

$$z_0 = r_0 (\cos \phi_0 + i \sin \phi_0),$$

and therefore $\quad w_0 = r_0^{\frac{1}{2}} (\cos \tfrac{1}{2} \phi_0 + i \sin \tfrac{1}{2} \phi_0),$

and if z describe a closed line once round the origin in the direction of increasing angles, then, on returning to z_0,

$$z = r_0 [\cos (\phi_0 + 2\pi) + i \sin (\phi_0 + 2\pi)],$$

and therefore $w = r_0^{\frac{1}{2}} [\cos (\tfrac{1}{2} \phi_0 + \pi) + i \sin (\tfrac{1}{2} \phi_0 + \pi)]$

$$= - w_0.$$

If the variable describe the closed line twice, or if it describe another closed line which winds round the origin twice, then the argument of z increases by 4π, therefore that of w by 2π, and consequently the function then assumes again its original value.

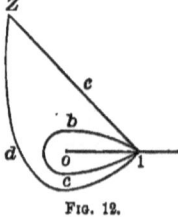

Fig. 12.

Now let the variable go from the point $z = 1$ to an arbitrary point Z, first along a line $1\,eZ$ (Fig. 12), which does not enclose the origin, and along which the angles ϕ increase. Along this path r and ϕ may acquire at Z the values R and θ, and w the value W, so that

$$W = R^{\frac{1}{2}} (\cos \tfrac{1}{2} \theta + i \sin \tfrac{1}{2} \theta).$$

But if the variable move upon the other side of the origin from 1 to Z along a line $1\,dZ$ not enclosing the origin, the angle ϕ decreases and acquires at Z the value $\theta - 2\pi$. Hence at Z in this case

$$z = R [\cos (2\pi - \theta) - i \sin (2\pi - \theta)],$$

and $\quad w = R^{\frac{1}{2}} [\cos (\pi - \tfrac{1}{2} \theta) - i \sin (\pi - \tfrac{1}{2} \theta)]$

$$= - W.$$

Finally, let z first describe a closed line $1bc1$ round the origin starting from 1, and next the line $1dZ$, then ϕ first increases from 0 to 2π and next decreases by the angle $2\pi - \theta$, so that ϕ acquires at Z the value $2\pi + \theta - 2\pi = \theta$; in this case w, after the description of the line $1bc1$, starts from 1 with the value -1 and acquires at Z the value $+W$ along $1dZ$.

Ex. 2. In the function
$$w = (z-1)\sqrt{z}$$
$z = 0$ is a branch-point, and this function behaves with respect to this point like the preceding. Let us consider therefore the point $z = 1$, for which likewise $w = 0$. Let the variable z describe round it a circle with radius r, starting from the point $a = 1 + r$ on the principal axis (Fig. 13). If we put
$$z - 1 = r(\cos\phi + i\sin\phi),$$
then $\quad w = r(\cos\phi + i\sin\phi)\sqrt{1 + r\cos\phi + ir\sin\phi}.$

Since r remains constant, and ϕ increases from 0 to 2π, the factor $r(\cos\phi + i\sin\phi)$ does not change its value. In order to study the behavior of the second factor, let

$$1 + r\cos\phi = \rho\cos\psi,$$
$$r\sin\phi = \rho\sin\psi;$$

then ρ denotes the straight line $\overline{oz}$, and ψ the inclination of the same to the principal axis, and

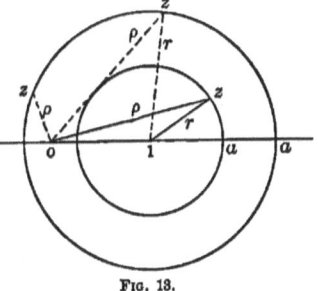

Fig. 13.

$$w = r(\cos\phi + i\sin\phi)\rho^{\frac{1}{2}}(\cos\tfrac{1}{2}\psi + i\sin\tfrac{1}{2}\psi).$$

Now, if the circle do not enclose the origin, ψ passes through a series of values commencing with 0 and ending with the value 0 again; hence w does not change its value. But if the circle be so large that the origin, which is a branch-point, also lies within it, ψ increases from 0 to 2π, and therefore in that case the original value $w = r\rho^{\frac{1}{2}}$ passes into $-r\rho^{\frac{1}{2}}$. The statement is therefore confirmed, that only the point $z = 0$ is a branch-point, and not the point $z = 1$.

We can consider the given function $(z-1)\sqrt{z}$ as derived from
$$w' = \sqrt{(z-1)(z-b)z}$$
by making $b = 1$. A line enclosing the point $z = 1$ can then be regarded as a line which at first enclosed the two points $z = 1$ and $z = b$, and in connection with which these two points were subsequently made to coincide. Now $z = 1$ and $z = b$, as well as $z = 0$, are branch-points of the function w'. A closed line which, starting from a point z_0, makes a circuit round both points 1 and b can be replaced by closed lines, each of which encloses only one of these points. If now w' start from z_0 with the value w_0', on encircling the point b it passes into $-w_0'$, and then on encircling the point 1, $-w_0'$ passes into w_0' again. The function returns therefore to z_0 with its original value. This continues to hold when b approaches the point 1, and when these branch-points coincide the common point obviously ceases to be a branch-point. It is evident that this may be generalized as follows: When once in connection with two branch-points only two function-values, and these two the same, pass mutually one into the other, these branch-points neutralize each other on coinciding, and there arises a point which is no longer a branch-point.

Ex. 3. Let $$w = \sqrt[3]{\frac{z-a}{z-b}},$$

in which a and b denote two complex constants. In this example we have two branch-points, $z = a$ and $z = b$. If we first let z describe a closed line round the point a starting from an arbitrary point z_0, but not enclosing the point b, and if we accordingly put
$$z - a = r(\cos\phi + i\sin\phi),$$
while
$$z_0 - a = r_0(\cos\phi_0 + i\sin\phi_0),$$
then the initial value of w, which may here be denoted by w_0, is

$$w_1 = \frac{r_0^{\frac{1}{3}}(\cos\frac{1}{3}\phi_0 + i\sin\frac{1}{3}\phi_0)}{[a - b + r_0(\cos\phi_0 + i\sin\phi_0)]^{\frac{1}{3}}}.$$

After the closed line is traversed once in the direction of increasing angles, ϕ_0 has increased by 2π, and hence the resulting value of w, which will be denoted by w_2, is

$$w_2 = \frac{r_0^{\frac{1}{3}}[\cos(\frac{1}{3}\phi_0 + \frac{2}{3}\pi) + i\sin(\frac{1}{3}\phi_0 + \frac{2}{3}\pi)]}{[a - b + r_0(\cos\phi_0 + i\sin\phi_0)]^{\frac{1}{3}}}.$$

Therein the denominator, and therefore the quantity $\sqrt[3]{z-b}$, cannot have changed its value, because for it $z = a$ is not a branch-point, but only $z = b$; therefore z has described a closed line which does not include the branch-point of this expression. Let

$$\cos\tfrac{2}{3}\pi + i\sin\tfrac{2}{3}\pi = \frac{-1 + i\sqrt{3}}{2} = \alpha,$$

so that α is a root of the equation $\alpha^3 = 1$; then, since

$$\cos(\tfrac{1}{3}\phi_0 + \tfrac{2}{3}\pi) + i\sin(\tfrac{1}{3}\phi_0 + \tfrac{2}{3}\pi)$$
$$= (\cos\tfrac{1}{3}\phi_0 + i\sin\tfrac{1}{3}\phi_0)(\cos\tfrac{2}{3}\pi + i\sin\tfrac{2}{3}\pi),$$

we can also write $\quad w_2 = \alpha w_1.$

Now let the variable again describe a closed line round the point a; then w leaves z_0 with the value $w_2 = \alpha w_1$, and therefore acquires, after the completion of the circuit, the value

$$w_3 = \alpha w_2 = \alpha^2 w_1.$$

After a third circuit w finally acquires the value $\alpha^3 w_1$, i.e., the original value w_1 again, since $\alpha^3 = 1$. If we had originally started from z_0 with the value w_2 instead of w_1, we should have obtained the values w_3 and w_1 after one and two circuits respectively; but if w_3 had been the original value, this would have changed into w_1 and w_2 successively.

Similar results are obtained when z is made to describe a closed line including only the point b. We then put

$$z - b = r(\cos\phi + i\sin\phi),$$

and let w start from z_0 with the value w_1, this value denoting the following expression:

$$w_1 = \frac{[b - a + r_0(\cos\phi_0 + i\sin\phi_0)]^{\frac{1}{3}}}{r_0^{\frac{1}{3}}(\cos\frac{1}{3}\phi_0 + i\sin\frac{1}{3}\phi_0)}.$$

After one circuit by z in the direction of increasing angles, the value of w becomes

$$\frac{[b - a + r_0(\cos\phi_0 + i\sin\phi_0)]^{\frac{1}{3}}}{r_0^{\frac{1}{3}}[\cos(\frac{1}{3}\phi_0 + \frac{2}{3}\pi) + i\sin(\frac{1}{3}\phi_0 + \frac{2}{3}\pi)]},$$

in which now the numerator cannot have changed its value, because its branch-point a has not been enclosed in the circuit. We therefore now obtain for w the value

$$\frac{w_1}{\alpha} = \alpha^2 w_1, \text{ i.e., the value } w_2.$$

After a second circuit we obtain

$$\frac{w_1}{\alpha^2} = \alpha w_1, \text{ i.e., } w_3;$$

finally, after a third circuit, the original value is restored, since

$$\frac{w_1}{\alpha^3} = w_1.$$

It is thus evident that the function-values for repeated circuits round a branch-point interchange in cyclical order. When z moves round the point a in the direction of increasing angles, the values

$$w_1, w_2, w_3$$

change after the first circuit respectively into

$$w_2, w_3, w_1,$$

and after the second circuit into

$$w_3, w_1, w_2;$$

after a third circuit therefore the original values

$$w_1, w_2, w_3$$

are restored. In like manner, for circuits round the point b in the direction of increasing angles, the values

$$w_1,\ w_2,\ w_3$$

pass into $\qquad w_3,\ w_1,\ w_2,$

and into $\qquad w_2,\ w_3,\ w_1,$

and acquire, after the third circuit, the original values

$$w_1,\ w_2,\ w_3.$$

Let us next inquire what takes place when z describes a closed line including both points, a and b. Such a line can always be deformed, without passing over one of these points, into another which consists of successive circuits round them (Fig. 11). Let then z first describe a circuit round the point b starting from z_0, return to z_0, and then describe a circuit round the point a. By this path w acquires, on the second return to z_0, the same value as when z describes a closed line round both branch-points (§ 9).

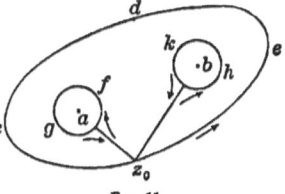

Fig. 11.

If w start from z_0 with the value w_1, it acquires the value $\dfrac{w_1}{a} = w_3$ after the circuit round b, and then after the circuit round a the value $aw_3 = w_1$; the function reverts therefore to its original value. If we consider in this connection, instead of the given function, the following:

$$w' = \sqrt[3]{(z-a)(z-b)},$$

in which, as is easily seen, the factor a is multiplied into the original function-value after each circuit round the point b; then w'_1 changes into $aw'_1 = w'_2$ on making the circuit round b, and on making the circuit round a, w'_2 changes into $aw'_2 = w'_3$. A circuit round both points therefore changes w'_1 into w'_3; hence a second circuit will change w'_3 into w'_2, and a third w'_2 into w'_1.

54 THEORY OF FUNCTIONS.

Ex. 4. The function
$$w = \sqrt[3]{\frac{z-a}{z-b}} + \sqrt{z-c},$$
which is the root of the equation of the sixth degree,
$$(z-b)^2 w^6 - 3(z-b)^2(z-c)w^4 - 2(z-a)(z-b)w^3$$
$$+ 3(z-b)^2(z-c)^2 w^2 - 6(z-a)(z-b)(z-c)w$$
$$+ (z-a)^2 - (z-b)^2(z-c)^3 = 0,$$

has the branch-points a, b, c. If, for sake of brevity, we substitute
$$\sqrt[3]{z-a} = t, \quad \sqrt[3]{z-b} = u, \quad \sqrt{z-c} = v,$$
and give to α the same meaning as in the preceding example, we can write the six function-values as follows:

$$w_1 = \frac{t}{u} + v, \qquad w_4 = \frac{t}{u} - v,$$
$$w_2 = \alpha \frac{t}{u} + v, \qquad w_5 = \alpha \frac{t}{u} - v,$$
$$w_3 = \alpha^2 \frac{t}{u} + v, \qquad w_6 = \alpha^2 \frac{t}{u} - v.$$

Let us first consider circuits of the variable round the point a; for these t passes into αt, $\alpha^2 t$, t, $\cdots$, while u and v remain unchanged; therefore

						$w_1\,w_2\,w_3$	$w_4\,w_5\,w_6$
change	after	the	first	circuit	into	$w_2\,w_3\,w_1$	$w_5\,w_6\,w_4$
"	"	"	second	"	"	$w_3\,w_1\,w_2$	$w_6\,w_4\,w_5$
"	"	"	third	"	"	$w_1\,w_2\,w_3$	$w_4\,w_5\,w_6$.

Round this branch-point, therefore, only the values w_1, w_2, w_3 permute by themselves, and w_4, w_5, w_6 by themselves.

For circuits round the point b, t and v remain unchanged, and u changes into αu, $\alpha^2 u$, u, $\cdots$. Therefore

						$w_1\,w_2\,w_3$	$w_4\,w_5\,w_6$
change	after	the	first	circuit	into	$w_3\,w_1\,w_2$	$w_6\,w_4\,w_5$
"	"	"	second	"	"	$w_2\,w_3\,w_1$	$w_5\,w_6\,w_4$
"	"	"	third	"	"	$w_1\,w_2\,w_3$	$w_4\,w_5\,w_6$.

Here the same function-values permute as for the point a, but in reverse sequence.

Finally, on making circuits round the point c, t and u are unchanged, and v changes into $-v$, $+v$, $\cdots$. Hence here

change after the first	circuit into	$w_1\,w_2\,w_3$		$w_4\,w_5\,w_6$
" " " second	" "	$w_4\,w_5\,w_6$		$w_1\,w_2\,w_3$
		$w_1\,w_2\,w_3$		$w_4\,w_5\,w_6$

In this example, therefore, we have first two branch-points, round which the three values w_1, w_2, w_3 permute in cyclical order, but never with one of the three remaining; likewise, w_4, w_5, w_6 permute in cyclical order here, but never pass into one of the first three values. We then have one more branch-point c, at which the three pairs w_1w_4, w_2w_5, w_3w_6, each by itself, interchange their values, without a value from one pair ever entering another.

If we let z describe a closed line including two branch-points, we can again replace such a one by two successive circuits, each round one point. If the points a and b be enclosed, we have the same condition as in the preceding example. We will therefore follow only circuits round a and c, and tabulate the results below.

Circuits.	Round a.	Round c.	Round Both.
1	w_1 changes into w_2	w_2 into w_5	w_1 into w_5
2	w_5 " " w_6	w_6 " w_3	w_5 " w_3
3	w_3 " " w_1	w_1 " w_4	w_3 " w_4
4	w_4 " " w_5	w_5 " w_2	w_4 " w_2
5	w_2 " " w_3	w_3 " w_6	w_2 " w_6
6	w_6 " " w_4	w_4 " w_1	w_6 " w_1

Therein w acquires its original value only after six consecutive circuits round the points a and c.

11. The preceding considerations show that, given a multiform function, we can pass continuously from one of the values which the function can assume for the same value of the variable to another, by assigning complex values to the variable and letting it pass through a series of continuously successive values, which ends with the same value with which it began (geometrically expressed, by letting the variable describe a closed line). It has been further shown that a definite and continuous series of values of the variable (a definite path), also, always leads to a definite function-value, except in the single case when the path of the variable leads through a branch-point, a case, however, which can always be obviated by letting the variable make an indefinitely small deviation in the vicinity of the branch-point.[1] This naturally suggests the desirability of avoiding the multiplicity of values of a multiform function, in order to be able to treat such a function as if it were uniform. According to the preceding explanations, it is necessary for this purpose only to do away with the multiplicity of paths which the variable can describe between two given points. Now Cauchy has already remarked that this could be effected, at least to a limited extent, by demarcating certain portions of the plane in which the variable z is supposed to be moving and not permitting the latter to cross the boundary of such a region. For, since a function, starting from a point z_0 of the variable, can assume different values at another point z_1, only when two paths described by the variable enclose a branch-point (§ 9), it is always easy to mark off a portion of the z-plane within which two such paths from z_0 to z_1 are not possible, or, by drawing certain lines which start from branch-points, and which are not to be crossed, to make such paths impossible. Within such a region the function remains uniform, since it acquires at each point z_1 only a single value along all paths. The function is then called *monodromic* (after Cauchy) or uniform, one-valued (after Riemann). Although this method is of great advantage, for instance, in the evalua-

---

[1] If we regard the position of the path of the variable, in case it lead through a branch-point, as the limiting position of a path not meeting the branch-point, then to this assumption corresponds again a definite function-value.

MULTIFORM FUNCTIONS.

tion of definite integrals, nevertheless by means of it only a definite region of values, or, as Riemann calls it, a definite *branch* of the multiform function, is separated from the rest and considered by itself. In order to be able to treat an algebraic function in its entirety and yet as if it were a uniform function, Riemann has devised another method, which will be set forth in the following.

Riemann assumes that, when a function is n-valued, when therefore to every value of the variable n values of the function correspond, the plane of z consists of n sheets or leaves (or that n such sheets are extended over the z-plane), which together form the region for the variable. To each point in each sheet corresponds only a single value of the function, and to the n points lying one immediately below another in all the n sheets correspond the n different values of the function which belong to the same value of z. Now at the branch-points, where several function-values, elsewhere different, become equal to one another, several of those sheets are connected, so that the particular branch-point is supposed to lie at the same time in all these connected sheets. The number of these sheets thus connected at a branch-point can be different at each branch-point, and is equal to the number of function values which change one into another in cyclical order for a circuit of the variable round the branch-point. In the last example of the preceding paragraph, wherein the function is six-valued, we shall assume the z-plane as consisting of six sheets. Round each of the branch-points a and b the values w_1, w_2, w_3, on the one hand, and w_4, w_5, w_6, on the other, change one into another; hence we assume that at each of these points the sheets 1, 2, 3, on the one hand, and the sheets 4, 5, 6, on the other, are connected. Round the point c, however, firstly, w_1 and w_4; secondly, w_2 and w_5; thirdly, w_3 and w_6, change respectively one into the other; hence at the point c, first the sheets 1 and 4, then the sheets 2 and 5, and finally the sheets 3 and 6 are connected. Now for the purpose of exhibiting the continuous change of one value of the function into another, so-called *branch-cuts*[1] are

[1] Sometimes called *cross-lines*. (Tr.)

introduced. These are quite arbitrary lines (except that one cannot intersect itself), which either pass from a branch-point to infinity, or join with each other two branch-points. We do not suppose the sheets to be connected along these branch-cuts as they naturally lie one above another, but as the function-values interchange round the respective branch-points. If, for instance, in the last example of the preceding paragraph, we draw a branch-cut from a to b (Fig. 14), we then, in making a circuit round the point a in the direction of increasing angles, connect the sheet 1 with the sheet 2 along the branch-cut, then 2 with 3, and finally 3 with 1 again. Let us call the right side of the branch-cut ab, that which an observer has on his right, when he stands at a and looks toward b. Then if z go from a point z_0 in sheet 1 (w with the value w_1) and make a circuit

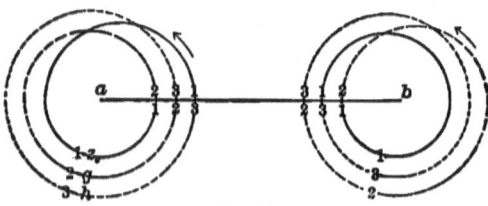

Fig. 14.

round the point a in the direction of increasing angles, on crossing the branch-cut from right to left, it passes from the first sheet into the second and is still in the latter when it returns to z_0, or rather to the point g lying immediately below z_0 in the second sheet, so that w acquires the value w_2. If the description be still continued, z passes, after crossing the branch-cut a second time from right to left, into the third sheet and is still in the same when it arrives at the point h situated in this sheet below z_0; w has now acquired the value w_3. Finally, when z crosses the branch-cut a third time, we assume that the right side of the third sheet is connected along the branch-cut with the left side of the first sheet through the second sheet, so that z crosses from the third sheet to the first sheet and then returns again to z_0.

Not until now is the line actually closed, and has w acquired again its original value. In Fig. 14 the lines are denoted by the numbers of the sheets in which they run, and, in addition, those in the second and third sheets are thickly dotted and thinly dotted respectively. The points z_0, g, h, which ought really to be one directly below another, are, for the sake of clearness, drawn side by side.

A similar course must be imagined in the case of all branch-points, and since from each such point a branch-cut starts, the variable cannot make a circuit round the branch-point without crossing the branch-cut, and thereby passing in succession into all those sheets which are connected at the branch-point. How the branch-cut is to be drawn in each case depends upon the function to be investigated, and can generally be chosen in different ways. In our example a and b may be connected by such a cut, because with circuits round the point b in the direction of increasing angles the function w_1 changes into w_3 and this into w_2 (Fig. 14), and hence at b the same sheets are connected as at a, and in the same way; namely, the right side of 1 with the left of 2, the right side of 2 with the left of 3, and the right side of 3 with the left of 1.[1]

Let us continue with this example and investigate the cir-

[1] If we wish to exhibit this method of representation by a model, a difficulty arises, first, because the sheets of the surface interpenetrate, and in the second place, because frequently at branch-points several sheets, which do not lie one immediately below another, must be supposed to be connected. But for the purpose of illustration, it is for the most part necessary only to be able to follow certain lines in their course through the different sheets of the surface. This can be easily effected in the following way: First cut in the sheets of paper placed one above another, which are to represent the surface, the branch-cuts, and then only at those places where a line is to pass over a branch-cut from one sheet into another join the respective sheets by pasting on strips of paper. Then we can always so contrive that, when the line is to return to the first sheet, from which it started, we have the necessary space left for the fastening of the strip of paper by means of which the return passage is effected. By these attached paper strips union of the separate sheets into one connected surface is accomplished; and it is then no longer necessary to connect the sheets with one another at the branch-points.

cuits discussed in the preceding paragraph round a and b, and round a and c. For a circuit round a and b the branch-cut is not crossed at all, so that z remains in the first sheet; in fact w resumes its original value at z_0 after such a circuit. (Cf. Ex. 3, § 10). To examine the circuit round the points a and c, let us draw from c to infinity a branch-cut, and let here the sheets of every pair 1, 4; 2, 5; 3, 6 pass respectively into each other.

For the passages of the function-values taking place here, we have found the following table (p. 55):

Circuits.	Round a.	Round c.	Round Both.
1	w_1 changes into w_2	w_2 into w_5	w_1 into w_5
2	w_5 " " w_6	w_6 " w_3	w_5 " w_3
3	w_3 " " w_1	w_1 " w_4	w_3 " w_4
4	w_4 " " w_5	w_5 " w_2	w_4 " w_2
5	w_2 " " w_3	w_3 " w_6	w_2 " w_6
6	w_6 " " w_4	w_4 " w_1	w_6 " w_1

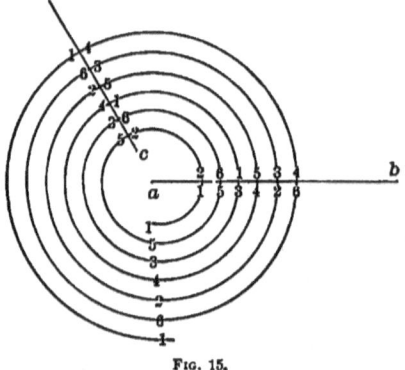

Fig. 15.

These passages are represented in Fig. 15, by designating each line by the number of the sheet in which it runs. The points properly lying below the initial point 1 are represented side by side, for the sake of clearness, and the last point 1 is to be supposed to coincide with the first.

This region for the variable z, consisting in our example of six sheets, forms a single connected surface, since the sheets are connected at the branch-points and pass into one another along the branch-cuts. In this surface w is a perfectly uniform function of the position in the surface, since it acquires the same value at every point of the latter, along whatever path the variable may arrive at the point. If z describe between two points two paths, which enclose a branch-point, then one of the two must necessarily cross a branch-cut and thereby pass into another sheet, so that the terminal points of the two paths are no longer to be considered coincident, but as two different points of the z-surface, at which different function-values occur. But if z describe an actually closed curve, *i.e.*, if the initial and the terminal points of the curve coincide in the same point of the same sheet, then also the function acquires its initial value. Only when the variable passes through a branch-point can it pass at will into any one of the sheets connected at that point, and in that case it remains undetermined which value the function assumes (§ 8).

12. Now in order to prove that we can also, in general, transform an n-valued function into a one-valued by means of an n-fold surface covering the z-plane, the single sheets of which are connected at the branch-points and along the branch-cuts in the manner explained above, we first assume the z-plane to be still single, and let the variable z, starting from an arbitrary point z_0, describe a closed line, which makes a circuit once round a single branch-point and does not pass through any other branch-point. At z_0 the function possesses n values; let us assume them to be written down in a certain sequence. Now after the variable has described the closed line and returned again to z_0, each of the above n function-values will have either passed into another or remained unchanged. Since the variable z is again at the point z_0, these new values of the function cannot differ from the former in their totality; but if we suppose them to be written down in the sequence in which they have arisen in succession from the former, they will occur in an arrangement different from the previous one.

Any one arrangement of n elements, however, can be derived from another arrangement by a series of cyclical permutations. By a cyclical permutation of the pth order we understand one in which from the existing n elements we take out p arbitrarily and in the place of the first of these put a second, in the place of the second a third, etc., and finally in the place of the pth the first. Such a cyclical permutation of the pth order has the property, that after p repetitions of it, and not sooner, the original arrangement is restored; for, since the place of each element is taken by another, that of the pth element, however, by the first, each element can reappear in its original place only after all the $p-1$ other elements have occupied the same place; then, however, each element actually returns to its original place. Now, to prove that each arrangement can be derived from another by a series of cyclical permutations, we assume that any one arrangement arises from another by substituting one element, say 3, for another, say 1. The place of 3 is then taken either by 1, in which case we have already a cyclical permutation of the second order, or by some other element, say 5. The place of the latter is then again taken either by the first, 1, in which case we have a cyclical permutation of the third order, or again by another which is different from those already employed, 1, 3, 5. Its place can be taken either by the first, whereby a cyclical permutation would be closed, or again by another; finally, however, the cyclical permutation must terminate, because altogether there is only a finite number of elements, and the first element 1 must be found in some place of the second arrangement. In this way then a series of elements is disposed of. If we next begin with some one of the elements not yet employed, we can repeat the former procedure until all the elements have been exhausted, and we thus obtain a certain number of cyclical permutations which, employed either successively or also simultaneously, produce the second arrangement from the first. If an element have not changed its place in the second arrangement, such a permanence can be regarded as a cyclical permutation of the first order. Let us illustrate the above by an example. Suppose the elements

$$1\quad 2\quad 3\quad 4\quad 5\quad 6\quad 7\quad 8\quad 9\quad 10\quad 11$$

have passed into the arrangement

$$3\quad 11\quad 5\quad 2\quad 7\quad 10\quad 1\quad 9\quad 6\quad 8\quad 4;$$

it is evident that the row
$$1\quad 3\quad 5\quad 7$$
has changed into $\quad 3\quad 5\quad 7\quad 1;$

these therefore form a cyclical permutation of the fourth order. If we next proceed from 2, it is evident that the row

$$2\quad 11\quad 4$$

changes into $\quad 11\quad 4\quad 2;$

therefore we have a second cyclical permutation, of the third order. The next element not yet employed is 6. Then the

row $\qquad\qquad 6\quad 10\quad 8\quad 9$

changes into $\qquad 10\quad 8\quad 9\quad 6,$

and we have a third cyclical permutation, of the fourth order. Now all 11 elements are exhausted, and consequently the second given arrangement is derived from the first by the three cyclical permutations obtained above.

If we now return to our function-values, it follows that, whatever arrangement of them may have arisen from the circuit of the variable round the branch-point, it can be produced by a series of cyclical permutations of the function-values. If the variable be made to describe a circuit round the same branch-point a second time, each function-value undergoes the same change that it experienced the first time. For this second circuit, therefore, the cycles remain the same as for the first, and so, too, for each subsequent circuit. Thus the values of the function (unless they all form a single cycle, which case can also occur; cf. Ex. 3, § 10) are divided at each branch-point into a series of cycles, so that in each cycle only certain definite values of the function can permute among themselves with the total exclusion of all values contained in another cycle (cf. Ex. 4, § 10).

If a single value of the function do not change for the circuit of the variable round the branch-point, the same can be regarded, as has been remarked above, as forming by itself a cycle of the first order. If the variable be now made to describe some quite arbitrary closed line, the latter can be deformed into a series of circuits round single branch-points (§ 9). Therefore the arrangement arising through this closed line can also be produced by the cyclical permutations occurring at the branch-points.

If, now, the z-surface be supposed to consist of n sheets, the preceding justifies the assumption that, at each branch-point, certain sets of sheets are supposed to be connected, which continue into one another along the branch-cuts in the way above stated. Then the variable, by describing a circuit round a branch-point, passes in succession into all those sheets belonging to the same group, and into none but those, and finally it returns into that sheet from which it started.

To find for an n-valued algebraic function the Riemann n-sheeted surface, let us first determine its branch-points and choose some definite value z_0 of z, which, however, must not itself be a branch-point. We must then let z describe a circuit in the single z-plane once round each separate branch-point, always starting from z_0 and returning again to it, and we must ascertain how the function-values occurring at z_0 are divided at each branch-point into the above-mentioned cycles, and how, within these, they permute with one another.[1]

---

[1] On this point cf. Puiseux, "Recherches sur les fonctions algébriques," *Liouville, Journ. de Math.* T. xv. (In the German by H. Fischer, *Puiseux's Untersuchungen über die algebraischen Funktionen.* Halle, 1861.) Königsberger, *Vorlesungen über die Theorie der ell. Funktionen.* Leipzig, 1874. I. S. 181.

This having been ascertained, if in the n-sheeted surface the points of the n-sheets which represent the value z_0 be designated in succession by $z_1^\circ, z_2^\circ, \cdots, z_n^\circ$, so that the subscript indicates the sheet in which the point is situated, then we can first arbitrarily distribute the values of the function at z_0 among the sheets; *i.e.*, we can assume in an arbitrary but definite way which of these values of the function shall belong to each of the points $z_1^\circ, z_2^\circ, \cdots, z_n^\circ$. We will denote these values in order by $w_1^\circ, w_2^\circ, \cdots, w_n^\circ$. Let us next draw from each branch-point a branch-cut to infinity, and for each of the latter let us so determine the connection of the sheets that it shall exactly correspond to the cycles previously ascertained. If, therefore, in the single z-plane, for a single circuit round a certain branch-point, w_a° be changed into w_β°, w_κ° into w_λ°, etc., then in the n-sheeted surface the connection of the sheets is so determined that for a single circuit round the same branch-point, the variable passes from z_a° to z_β°, from z_κ° to z_λ°, etc. If a single function-value w_μ° suffer no change thereby, the corresponding sheet μ is connected with no other sheet, so that the sheet μ continues into itself along the branch-cut, it being therefore unnecessary to draw the branch-cut in this sheet.

If, as has been assumed, a branch-cut extend from each branch-point to infinity, the connection of the sheets can be determined at each branch-point independently of the others. This, however, does not exclude the possibility of sometimes connecting two branch-points by a branch-cut, or of making several branch-cuts extend from one branch-point; but this may occur only when the previously ascertained way in which the function-values permute at the respective branch-points permits such an arrangement. Thus, in the example of a six-valued function considered in the preceding paragraph, a branch-cut can be drawn from each of the three branch-points a, b, c to infinity; but the way in which the values of the function permute round a and b also admits of a and b being connected by a branch-cut.

These determinations having once been established, the

function-values for each value of z are distributed among the n sheets in a definite way. To prove this, since in the single z-plane the different values which the function can have at one and the same point z are produced only by the different paths to z, it is only necessary to show that, if in the n-sheeted surface starting from a certain point, say z_1°, and with the definite value w_1°, we reach the same arbitrarily chosen point z_λ along any two different paths, we are always led to the same function-value (by the two paths). In this proof different cases must be distinguished.

(1) Let us first assume that the terminal point of a path starting from z_1° is one of the points representing the value z_0, say z_λ°. Then the corresponding path in the single z-plane forms a closed line. This can be deformed into a series of (closed) circuits round single branch-points without changing the final value of the function. Corresponding to this in the n-sheeted surface, the variable also makes circuits round single branch-points and after each circuit goes to whichever of the points $z_1^\circ, z_2^\circ, \ldots, z_n^\circ$ it can reach. The points at which it arrives in this manner may be designated in order by $z_1^\circ, z_a^\circ, z_\beta^\circ, \ldots, z_\kappa^\circ, z_\lambda^\circ$. According to the principles established above concerning the arrangement of the function-values, and since here only single circuits round any one of the branch-points are taken into consideration, it follows that w assumes in succession the values $w_1^\circ, w_a^\circ, w_\beta^\circ, \ldots, w_\kappa^\circ, w_\lambda^\circ$. Now a similar deformation can be made in the case of any other path starting from z_1° and terminating in z_λ°, although then the variable will generally pass into other sheets than before. Let us assume that the circuits lead the variable in succession from z_1° to $z_a^\circ, z_b^\circ, \ldots, z_k^\circ$, and finally to z_λ°, then the corresponding series of function-values is $w_1^\circ, w_a^\circ, w_b^\circ, \ldots, w_k^\circ$, and since the last circuit, according to the assumption, leads from z_k° to z_λ°, w_k° must finally also change into w_λ°. This may be illustrated by an example.

In the six-sheeted surface represented in § 11 we can reach for instance z_3° from z_1° by two circuits round the point a, by which we come first from z_1° to z_2°, and then to z_3°. But among others we can also choose the following way: from z_1° round

a to $z_2°$, then round c to $z_5°$, then round a to $z_6°$, and finally round c to $z_3°$. For the first path w assumes the values $w_1°$, $w_2°$, $w_3°$; for the second, $w_1°$, $w_2°$, $w_5°$, $w_6°$, $w_3°$, but the final value is the same for both paths. It must be emphasized as a special case that, if the variable return by any one path to its starting-point $z_1°$, the function also resumes the original value $w_1°$.

(2) We next consider two paths A and B connecting the same points and running entirely in one and the same sheet; therefore, if for instance, we assume again $z_1°$ as the starting-point, entirely in the sheet 1. Then, if the two paths enclose no branch-point, no special discussion is required, since the corresponding paths in the single z-plane also enclose no branch-point, and therefore lead to the same value of the function (§ 9). It is to be noted that this case can occur, if the paths considered enclose branch-points, only when from neither of the enclosed branch-points a branch-cut extends to infinity; for, otherwise, at least one of the paths would have to cross the branch-cut and could not run entirely in the same sheet. The case under consideration can therefore only occur when several branch-points are enclosed by the two paths, and when each branch-cut connects two branch-points with each other. If then, in the single z-plane, we let circuits round all the enclosed branch-points precede one of the paths, say B, we obtain a new path C leading to the same function-value as does A. But if these circuits be made in the n-sheeted surface, they again lead back to $z_1°$, because each branch-cut connects two branch-points, and therefore, for the successive circuits round the latter, the branch-cut must be crossed twice in opposite directions; we thus always come back to the sheet 1, and therefore finally also to $z_1°$. Then, according to what has been proved in the preceding case, the function also acquires again at $z_1°$ the value $w_1°$. Since, however, the path C, which consists of the circuits and the path B, leads to the same value as does A, and since the function starts on the path B with the value $w_1°$, therefore this alone must also lead to the same value as does A.

(3) Finally, let us assume as the terminal point of the paths to be examined any point z_λ lying in any arbitrary sheet λ. Let

the initial point, as before, be z_1°. For such a path we can, in the first place, without changing the final function-value, substitute another path, which first leads to z_λ° and then, running entirely in the sheet λ, to z_λ; for, in the corresponding paths in the single z-plane, the terminal portion of the second can always be so chosen that this can be deformed into the first without necessarily crossing a branch-point. If we now make the same deformation in the two different paths, both first lead to z_λ°; here the function according to (1) acquires along both paths the value w_λ°. The portions of the two paths still remaining run entirely in the sheet λ, commence at the same point z_λ° and with the same value of the function w_λ°; therefore, both according to (2) also lead to the same value of the function at z_λ.

We have thus proved that after the above determinations, chosen quite arbitrarily, have been made, the function acquires at each point of the surface a definite value independent of the path and becomes a one-valued function of the position in the surface. Thereby we have removed the multiformity of algebraic functions,[1] and in what follows we shall now always assume that the region of the variable consists of as many sheets as are required to change the multiform function under consideration into a uniform one, and we shall consider two points as identical, only when they also belong to the same sheet of the surface.

Accordingly we shall call a line actually closed, only when its initial and terminal points coincide at the same point of the same sheet. On the other hand, should a line end in a point situated in another sheet above or below the initial point, we shall sometimes call such a line apparently closed.

13. To the above considerations let us add a few remarks. In crossing a branch-cut, one sheet is continued into another,

---

[1] Though we can also ascribe branch-points to such functions as $\log z$, $\tan^{-1}z$, etc., we should then be obliged to assume that an infinite number of sheets of the surface are connected in a branch-point. For this reason the functions mentioned will be considered later from the point of view of functions defined by integrals.

as has been set forth above, in such a manner that, when the variable moves in it, the function changes continuously. It follows from this, which is to be well heeded, that the function in the same sheet must always have different values on the two sides of a branch-cut. Let us assume, for example, that a sheet κ is continued beyond a branch-cut into another sheet λ, and let z_κ and z_λ be two points representing the same value of z, and lying in κ and λ respectively and infinitely near the branch-cut. Further, let z'_κ be the point which lies in κ on the other side of the branch-cut directly opposite z_κ and infinitely near the same. Then the variable, in order to pass from z_κ to z_λ, must first move round the branch-point to z'_κ, from which point it immediately comes to z_λ by crossing the branch-cut. Accordingly z'_κ and z_λ are in continuous succession; z_κ and z_λ, however, are not. If $w_\kappa, w'_\kappa, w_\lambda$ denote the function-values corresponding to $z_\kappa, z'_\kappa, z_\lambda$ respectively, then w_λ is continuous with w'_κ, but not with w_κ, and if we disregard the infinitely small difference between w_λ and w'_κ, we can say $w'_\kappa = w_\lambda$; but since w_λ is different from w_κ, w_κ and w'_κ are also different from each other. Take, for instance, the function $w = \sqrt{z}$. The surface then consists of two sheets, which are connected at the branch-point $z = 0$. Here $w_\lambda = - w_\kappa$ (cf. § 10, Ex. 1); therefore $w'_\kappa = - w_\kappa$. In this example, therefore, the values of the function in the same sheet have opposite signs on the opposite sides of the branch-cut.

Riemann calls the branch-points also *winding-points*, because the surface winds round such a point like a screw surface of infinitely near threads. Then, if only two sheets of the surface be connected at such a point, it is called a *simple branch-point*, or a *winding-point of the first order;* if, however, n sheets of the surface be connected at it, it is called a *branch-point* or *winding-point of the $(n-1)th$ order*. Now, for many investigations, it is important to show that a winding-point of the $(n-1)$th order may be regarded as one at which $n-1$ simple branch-points have coincided. If we assume, for example, $n = 5$, then at a branch-point in which 5 sheets are connected, the variable passes after each circuit into the next following

70   THEORY OF FUNCTIONS.

sheet, and a curve must make 5 circuits round a branch-point before it arrives again in the first and becomes closed. By this property such a point is characterized. But the same also takes place if we assume 4 simple branch-points a, b, c, d in which the following sheets are connected in succession:

at a b c d
      1 and 2   1 and 3   1 and 4   1 and 5.

In Fig. 16, aa', bb', cc', dd' are the branch-cuts, and the figures refer to the numbers of the sheets in which the lines run. If the curve, starting from z_0 (Fig. 16), cross the section aa', it passes from 1 into 2 and remains in 2 for the entire circuit, because this sheet is not connected with another at any of the points b, c, d. Thus, for the first circuit the curve passes from 1 into 2. If aa' be crossed a second time, the curve passes from 2 into 1, and then at bb' from 1 into 3. Then, however, it remains in 3 until its return to z_0, the second circuit therefore carrying it into 3. Only at bb' it again passes from 3 into 1, and then at cc' from 1 into 4. In this way each new circuit carries the curve into the next following sheet; therefore, after the fifth circuit the curve returns into the first sheet and becomes closed. It is thus seen that the passages take place here in the same way as in the case of a winding-point of the fourth order. Therefore, by making the four simple branch-points, as well as the branch-cuts, approach one another and finally coincide, we obtain a branch-point of the fourth order. This simple example shows at the same time that the number of circuits which a curve must make round a region in order to become closed, exceeds by 1 the number of the simple branch-points contained in this region, since the winding-point of the

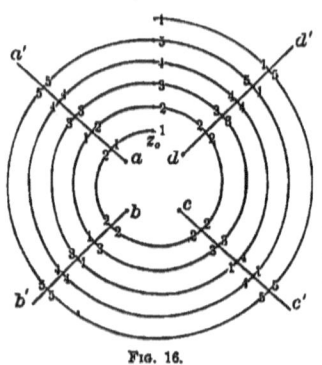

Fig. 16.

fourth order is equivalent to four simple branch-points. It will be shown later that this relation holds generally.

14. For the complete treatment of the algebraic functions it is still requisite for us to take into consideration infinitely large values of the variable z. In the plane in which z is moving, this variable can move from any given point, *e.g.*, from the origin, in any direction to infinity. But if we now suppose the plane to be closed at infinity, like a sphere with an infinitely large radius, we can imagine that all those directions extending to infinity meet again in a definite point of the sphere, and accordingly the value $z = \infty$ can then be represented by a definite point on the spherical surface. The same representation may be obtained by conceiving that the z-plane is tangent at the origin to a sphere of arbitrary radius. Let us suppose the point of tangency to be the north pole of the sphere. Then any point z of the plane can be projected on the surface of the sphere by drawing a straight line from the south pole s of the sphere to the point z, and cutting with this line the surface of the sphere. But if the infinitely distant points of the z-plane be projected in this manner on the surface of the sphere, the projections all fall in the point s, by which point therefore the value $z = \infty$ is represented in that case.

If, now, the z-plane consist of n-sheets, the spherical surface can be supposed also to consist of n-sheets, and we can assume that the points of the n-sheets representing the value $z = \infty$ lie directly one above another. Then it is also conceivable that several sheets are connected at the point ∞, and that the latter is a branch-point. Given a function $w = f(z)$, in order to decide whether $z = \infty$ is a branch-point, we need only substitute $z = \dfrac{1}{u}$. If, then, $f(z)$ change into $\phi(u)$, each branch-point $z = a$ of $f(z)$ furnishes for $\phi(u)$ a branch-point $u = \dfrac{1}{a}$, and, conversely, each branch-point $u = b$ of $\phi(u)$ furnishes a branch-point $z = \dfrac{1}{b}$ for $f(z)$; therefore $z = \infty$ is or is not a branch-point of $f(z)$, according as $u = 0$ is or is not a branch-point of $\phi(u)$. We can also let the variable z describe a circuit round the point ∞ on the surface of the sphere, and we can ascertain whether or not $f(z)$ thereby undergoes a change in value, and what the nature of this change is, if for this purpose we examine the function $\phi(u)$, while the variable u describes a circuit round the point $u = 0$.

In the case of a surface closed at infinity, a branch-cut can no longer be drawn extending indefinitely to infinity, but if such a cut extend to infinity, it now terminates in the definite point $z = \infty$. Thence arises a difficulty which has to be removed (which, however, may be shown to be only apparent). For, let $a, b, c, \cdots$ be the finite branch-points of a given function $f(z)$, and let us assume that these points are connected by means of branch-cuts with the point $z = \infty$. If, now, the values of the function $w_1°, w_2°, \cdots, w_n°$, occurring for any definite value z_0, be distributed arbitrarily among the points $z_1°, z_2°, \cdots, z_n°$, as has been set forth on page 65, and if the connection of the sheets along the branch-cuts be determined in accordance with the character of the function $f(z)$, then nothing arbitrary must be assumed for the point $z = \infty$, but the manner in which the sheets are connected along the branch-cuts which end in this point is already determined by the former assumptions. The question, however, then arises whether this actually conforms to the nature of the function $f(z)$, i.e., whether thereby that change (or eventually non-change) of value is produced which $f(z)$ really experiences for a circuit round the point $z = \infty$. If this were not the case, it would be impossible so to construct the Riemann spherical surface that the given function should be changed into a one-valued one without neglecting the infinitely large value of z.

But, now in the simple spherical surface a closed line Z, which encloses the point $z = \infty$ and no other branch-point, is at the same time one which encloses all finite branch-points $a, b, c, \cdots$. Since the latter can be resolved into closed lines which enclose the branch-points $a, b, c, \cdots$, singly, the same change of values takes place for $f(z)$ in describing the line Z as if the points $a, b, c, \cdots$ were enclosed singly in succession.

MULTIFORM FUNCTIONS.

This, however, in the n-sheeted surface, as has been shown above, is at the same time the change which also occurs when the branch-cuts leading from $a, b, c, \ldots$ to infinity are crossed in succession. Accordingly, there arises, in fact, no contradiction, but it is always possible to represent uniformly an algebraic function by a many-sheeted Riemann spherical surface, without neglecting the infinitely large value of z.

For the purpose of illustrating the above, we shall introduce a few examples, in which, for the sake of brevity, the same designations will be used as have been employed in § 10 and § 11.

Ex. 1. The function already considered

$$f(z) = \sqrt[3]{\frac{z-a}{z-b}} + \sqrt{z-c}$$

changes by the substitution $z = \dfrac{1}{u}$ into

$$\phi(u) = \sqrt[3]{\frac{1-au}{1-bu}} + \frac{\sqrt{1-cu}}{\sqrt{u}};$$

hence $u = 0$, and therefore also $z = \infty$, is a branch-point, and it is evident that at this point the same sheets are connected as at the point c. We shall therefore draw one branch-cut from a to b, and a second from c to infinity. But it is also possible to draw three branch-cuts from a, b, c to infinity, as we have done in the general treatment of this subject. From the considerations made in Ex. 4 of § 10, it follows that the passages along the branch-cuts are as follows:

$$\text{along } a\infty \cdots \frac{1\;2\;3\;4\;5\;6}{2\;3\;1\;5\;6\;4}$$

$$\text{`` } b\infty \cdots \frac{1\;2\;3\;4\;5\;6}{3\;1\;2\;6\;4\;5}$$

$$\text{`` } c\infty \cdots \frac{1\;2\;3\;4\;5\;6}{4\;5\;6\;1\;2\;3}$$

If therefore for a single circuit round the point ∞ these three cuts be crossed successively, we then pass in succession first from 1 to 2, then to 1, and finally to 4, in accordance with what should really occur.

Ex. 2. $\quad f(z) = \sqrt[3]{\dfrac{(z-a)(z-b)}{z^2}}$

changes into $\quad \phi(u) = \sqrt[3]{(1-au)(1-bu)}$;

therefore $z = \infty$ is not a branch-point, but only the points 0, a and b.

From each of these points can be drawn a branch-cut to infinity. But if the surface be assumed to be closed at infinity, then the three branch-cuts meet at the point ∞ (Fig. 17). The sheets of the surface pass into one another along the part $a\,\infty$ in a way different from that along the part $b\,\infty$, namely, as the numbers indicate in the figure. Round the branch-point 0, in the direction of increasing angles, $f(z)$ changes into

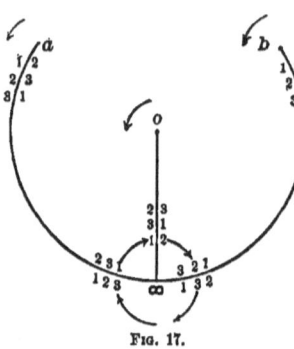

Fig. 17.

$$\dfrac{f(z)}{a^2} = af(z);$$

therefore 1 into 2, and hence also 2 into 3 and 3 into 1. If we now describe a circuit round the point ∞, then, on crossing $0\,\infty$, 1 changes into 2, and on crossing $b\,\infty$, 2 into 3, and finally on crossing $a\,\infty$, 3 into 1. Here therefore after the first circuit we return to the first sheet, the function does not change its value, and thus the point ∞ is really not a branch-point.

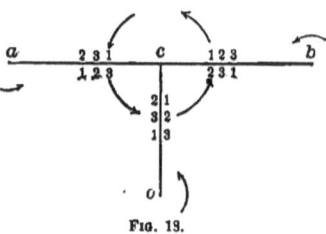

Fig. 18.

It would also have been possible in this case to connect the points a and b by a branch-cut drawn in the finite part of the

MULTIFORM FUNCTIONS. 75

surface (Fig. 18). But then there must be given on this line a point c, at which separation takes place, so that along ac the sheets are connected in a different way from that along bc. If then the second branch-cut be drawn from 0 to c, then for the circuit round the point c the function remains unchanged, so that c is not a branch-point. The matter can here be considered in a manner analogous to the treatment of the second example of § 10, namely, as if the point c had arisen from the coincidence of three branch-points d, e, f, which had mutually neutralized one another, so that the given function may be supposed to have been derived from the following

$$\sqrt[3]{\frac{z-a}{z-d} \cdot \frac{z-b}{z-e} \cdot \frac{(z-f)^2}{z^2}}$$

by putting $\qquad d = e = f = c.$

The branch-cuts can here also be chosen in a third way by drawing one from a to 0, and another from 0 to b.

Ex. 3. The function
$$f(z) = \sqrt[3]{(z-a)(z-b)}$$
changes into
$$\phi(u) = \sqrt[3]{\frac{(1-au)(1-bu)}{u^2}};$$

therefore $z = \infty$ is a branch-point. We can here draw a branch-cut from a to ∞, and another one from b to ∞ (Fig. 19), and connect the sheets as indicated in the figure. Then a circuit round the point ∞ leads first across $b\infty$ from 1 to 2, and then across $a\infty$ from 2 to 3; thus one circuit leads from 1 to 3, so that the function changes and $z = \infty$ is actually a branch-point. It is to be noted that if the direction of motion here, too, be that of increasing

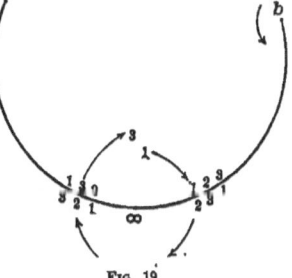

FIG. 19.

angles, the circuit, viewed from ∞, must be made in the opposite direction. For, if we put

$$u = r(\cos\phi - i\sin\phi),$$

it follows that

$$z = \frac{1}{r}(\cos\phi + i\sin\phi).$$

Therefore, if u describe a circle round the origin with a small radius and in the direction of decreasing angles, then z describes a circle with a large radius and in the direction of increasing angles. In this case $\phi(u)$, for one circuit, changes into $a^2\phi(u)$, and consequently $f(z)$ into $a^2 f(z)$; i.e., we pass from 1 to 3, as indicated in the figure.

We can here also connect the points a and b by a branch-cut running in the finite part of the plane, assume on it a point of separation c, and draw from this another branch-cut to ∞ (Fig. 20). The function then does not change for a circuit round the point c.

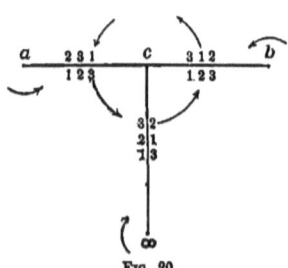

Fig. 20.

Ex. 4. Let us next take up the example previously given on page 38, in which the function w is defined by the cubic equation

$$w^3 - w + z = 0.$$

By letting here as above

$$p = \sqrt[3]{\tfrac{1}{2}(-z - \sqrt{z^2 - \tfrac{4}{27}})}, \quad q = \sqrt[3]{\tfrac{1}{2}(-z + \sqrt{z^2 - \tfrac{4}{27}})},$$

whereby
$$pq = \tfrac{1}{3}, \tag{1}$$

the three values of the function are expressed by

$$w_1 = p + q,$$
$$w_2 = ap + a^2 q,$$
$$w_3 = a^2 p + aq.$$

We have here first the two branch-points $z = \dfrac{2}{\sqrt{27}}$ and $z = -\dfrac{2}{\sqrt{27}}$, at each of which two sheets of the surface are connected; and next $z = \infty$ is also a branch-point, since by letting $z = \dfrac{1}{u}$, we get

$$p = \sqrt[3]{\dfrac{-1-\sqrt{1-\dfrac{4u^2}{27}}}{2u}};$$

for $u = 0$, p is thus $= \infty$, and therefore by (1) $q = 0$. Accordingly, at $z = \infty$ all three values of the function become infinitely large, so that all three sheets are connected. Although, at first sight, it seems as if the first two branch-points must have exactly the same relation, so that at both the same sheets are connected, yet this is not the case. To see this, it is only necessary to follow the real values of z, since the first two branch-points are real and the point $z = \infty$ can also be assumed on the principal axis; and the expressions for the roots w must be reduced to the forms which are given to them in the irreducible case of the cubic equation. We write therefore

$$p = \sqrt[3]{\tfrac{1}{2}\left(-z - \dfrac{2}{\sqrt{27}}\sqrt{\dfrac{27 z^2}{4} - 1}\right)},$$

and let

$$z = \dfrac{2}{\sqrt{27}} \cos 3v.$$

Then we get

$$p = \sqrt[3]{-\dfrac{1}{\sqrt{27}}(\cos 3v + i \sin 3v)} = -\sqrt[3]{\tfrac{1}{3}}(\cos v + i \sin v);$$

the three values of p therefore are

$$p = -\sqrt[3]{\tfrac{1}{3}}\, e^{iv}, \ = -a\sqrt[3]{\tfrac{1}{3}}\, e^{iv}, \ = -a^2\sqrt[3]{\tfrac{1}{3}}\, e^{iv},$$

and, since q is always $= \dfrac{1}{3p}$, the corresponding values of q are

$$q = -\sqrt[3]{\tfrac{1}{3}}\, e^{-iv}, \ = -a^2\sqrt[3]{\tfrac{1}{3}}\, e^{-iv}, \ = -a\sqrt[3]{\tfrac{1}{3}}\, e^{-iv}.$$

Since, moreover, $a = e^{\frac{2i\pi}{3}}$, $a^2 = e^{-\frac{2i\pi}{3}}$ can be substituted, we get

$$w_1 = -\sqrt[3]{\tfrac{1}{3}}\left(e^{iv} + e^{-iv}\right),$$

$$w_2 = -\sqrt[3]{\tfrac{1}{3}}\left(e^{i\left(v+\frac{2\pi}{3}\right)} + e^{-i\left(v+\frac{2\pi}{3}\right)}\right),$$

$$w_3 = -\sqrt[3]{\tfrac{1}{3}}\left(e^{i\left(v-\frac{2\pi}{3}\right)} + e^{-i\left(v-\frac{2\pi}{3}\right)}\right);$$

or

$$w_1 = -2\sqrt[3]{\tfrac{1}{3}}\cos v,$$

$$w_2 = -2\sqrt[3]{\tfrac{1}{3}}\cos\left(v + \frac{2\pi}{3}\right),$$

$$w_3 = -2\sqrt[3]{\tfrac{1}{3}}\cos\left(v - \frac{2\pi}{3}\right).$$

As long as v remains real, z passes through the real values numerically less than $\frac{2}{\sqrt{27}}$; the point z therefore describes the distance between the two simple branch-points; for pure imaginary values of v, z becomes numerically greater than $\frac{2}{\sqrt{27}}$, and only for complex values of v does z assume imaginary values. For our purpose therefore only the real values of v need be considered. In this it is to be noted, however, that z is introduced as a periodical function of v. Therefore, if we wish to have something definite and make the variable z describe the distance between the two simple branch-points only once, we must choose a definite period and consider this alone. Let us then assume, in order that z may pass through the real values $+\frac{2}{\sqrt{27}}$ to $-\frac{2}{\sqrt{72}}$, that $3v$ moves from 0 to π, and therefore v from 0 to $\frac{\pi}{3}$. We then obtain the following corresponding values of v, z, w_1, w_2, w_3:

v	z	w_1	w_2	w_3
0	$+\dfrac{2}{\sqrt{27}}$	$-2\sqrt[3]{\tfrac{1}{3}}$	$+\sqrt[3]{\tfrac{1}{3}}$	$+\sqrt[3]{\tfrac{1}{3}}$
$\dfrac{\pi}{6}$	0	-1	$+1$	0
$\dfrac{\pi}{3}$	$-\dfrac{2}{\sqrt{27}}$	$-\sqrt[3]{\tfrac{1}{3}}$	$+2\sqrt[3]{\tfrac{1}{3}}$	$-\sqrt[3]{\tfrac{1}{3}}$

In calculating them, to avoid all ambiguity at the branch-points, we must start from the value $v = \dfrac{\pi}{6}$ and make it first decrease to 0 and then increase to $\dfrac{\pi}{3}$.

If, for the sake of brevity, the branch-points $+\dfrac{2}{\sqrt{27}}$ and $-\dfrac{2}{\sqrt{27}}$ be denoted by e and e', it is seen that, though according to the chosen period of v the two values of the function w_2 and w_3 become equal at e, yet after z has arrived at e', this does not again take place, but now w_1 and w_3 become equal. Accordingly, at e the sheets 2 and 3, but at e' the sheets 1 and 3, must be assumed to be connected.

If, according to the above table, the three values -1, $+1$, 0 of w occurring for $z = 0$ be distributed consecutively among the sheets 1, 2, 3, then, for a circuit round the point e, the values $+1$ and 0 interchange; while for a circuit round the point e' the values -1 and 0 interchange, which is also confirmed by a direct investigation of these circuits. Accordingly, if the branch-cuts $e\infty$ and $e'\infty$ be drawn, the continuation of the sheets, in crossing them, must be assumed as follows:

$$\text{along } e\infty \cdots \dfrac{1\ 2\ 3}{1\ 3\ 2}$$

$$\text{`` } e'\infty \cdots \dfrac{1\ 2\ 3}{3\ 2\ 1}.$$

At the point ∞, then, all three sheets are connected, into which we pass successively if we describe a circuit round this point.

80   THEORY OF FUNCTIONS.

15. If w denote a multiform function of z, but W a rational function of w and z (or also of w alone), then the z-surface for the function W is constructed just as is that for the function w. For, let w_κ and w_λ denote any two values of w belonging to the same z, and W_κ and W_λ the corresponding values of W, then W_κ must change into W_λ whenever w_κ changes into w_λ, since to each pair of values of z and w corresponds only a single value of W. The passages of the W-values depend therefore upon the circuit described by z in the same manner as do the w-values.

Therefore the z-surface has the same branch-points and branch-cuts for W as for w, and at each branch-point the same sheets are connected. For this reason Riemann calls all rational functions of w and z a system of *like-branched* functions.

SECTION IV.

INTEGRALS WITH COMPLEX VARIABLES.

16. The definite integral of a function of a complex variable can be defined in exactly the same manner as is that of a function of a real variable.

Let z_0 and z be any two complex values of the variable z. Let the points which represent these values be connected by an arbitrary continuous line, and assume on it a series of intermediate points, which correspond to the values $z_1, z_2, \cdots, z_n$ of the variable. If, further, $f(z)$ be a function of z which at no point of the above line tends towards infinity, and if we form the sum of the products,

$$f(z_0)(z_1 - z_0) + f(z_1)(z_2 - z_1) + \cdots + f(z_n)(z - z_n),$$

then the limit of this expression, when the number of the intermediate values between z_0 and z along the arbitrary line

INTEGRALS WITH COMPLEX VARIABLES. 81

increases indefinitely, and when therefore the differences $z_1 - z_0$, $z_2 - z_1$, etc., diminish indefinitely, is the definite integral between the limits z_0 and z; therefore

$$\int_{z_0}^{z} f(z)\,dz = \lim\, [f(z_0)(z_1 - z_0) + f(z_1)(z_2 - z_1) + \cdots$$
$$+ f(z_n)(z - z_n)]. \quad (1)$$

It is obvious that this definition does not essentially differ from that usually given for real variables. One difference, however, consists in this: that, in accordance with the nature of a complex variable, the path described between the lower and upper limits, i.e., the series of intermediate values, is not a prescribed one, but can be formed by means of any continuous line. Upon the nature of this line, which is called the *path of integration*, the integral is in general absolutely dependent. It is easy to show that, if $f(z)$ do not become infinite at any point of a path of integration, the integral taken along this path has also a finite value. For, since (§ 2, 1) the modulus of a sum is less than the sum of the moduli of the single terms, it follows from (1) that

$$\mathrm{mod}\int_{z_0}^{z} f(z)\,dz < \lim\,\{\mathrm{mod}\,[f(z_0)(z_1 - z_0)]$$
$$+ \mathrm{mod}\,[f(z_1)(z_2 - z_1)] + \cdots + \mathrm{mod}\,[f(z_n)(z - z_n)]\}.$$

But if M denote the greatest of the values acquired by the modulus of $f(z)$, while z describes the path of integration, this value according to the assumption being finite, then the right side becomes still greater if M be put in place of the moduli of the single function-values $f(z_0)$, $f(z_1) \cdots$. Therefore, the modulus of a product being equal to the product of the moduli of the factors (§ 2, 3), we get

$$\mathrm{mod}\int_{z_0}^{z} f(z)\,dz < M \cdot \lim\,\{\mathrm{mod}\,(z_1 - z_0) + \mathrm{mod}\,(z_2 - z_1) + \cdots$$
$$+ \mathrm{mod}\,(z - z_n)\}.$$

In this the moduli of the differences $z_1 - z_0$, $z_2 - z_1 \cdots$ represent the lengths of the chords $\overline{z_0 z_1}$, $\overline{z_1 z_2}$, $\cdots$. In passing to the limit, therefore, the sum of these moduli approaches the length L of the path of integration; accordingly

$$\operatorname{mod} \int_{z_0}^{z} f(z)\, dz < ML,$$

and has a finite value, if the path of integration have a finite length.[1]

From this definition follow immediately the two following propositions: —

1. If z_k denote any value of the variable along its path, then

$$\int_{z_0}^{z} f(z)\, dz = \int_{z_0}^{z_k} f(z)\, dz + \int_{z_k}^{z} f(z)\, dz.$$

2. Also $\quad \int_{z}^{z_0} f(z)\, dz = - \int_{z_0}^{z} f(z)\, dz,$

i.e., if the variable describe the path which represents a continuous succession of its values in the opposite direction, the integral assumes the opposite sign.

It can further be shown that, whatever may be the path of integration, the integral

$$w = \int_{z_0}^{z} f(\zeta)\, d\zeta$$

is always a function of the upper limit z, when the lower limit z_0 is assumed to be constant. Let

$$z_0 = x_0 + iy_0, \quad z = x + iy,$$
$$\zeta = \xi + i\eta;$$

then we have $\quad w = \int_{x_0 + iy_0}^{x + iy} f(\xi + i\eta)\, (d\xi + id\eta).$

This integral breaks up into two parts, so that in the first ξ, and in the second η, is the variable of integration. Given now a definite path of integration, then by virtue of it η is a function of ξ, and ξ a function of η; let

$$\eta = \phi(\xi), \quad \xi = \psi(\eta).$$

[1] Königsberger, *Vorlesungen über die Theorie der ellipt. Funkt.*, I. S. 63.

INTEGRALS WITH COMPLEX VARIABLES.

If these be introduced, then, since ξ passes through all values from x_0 to x, and η at the same time through all values from y_0 to y corresponding to ξ in virtue of the path of integration,

$$w = \int_{x_0}^{x} f[\xi + i\phi(\xi)] d\xi + i \int_{y_0}^{y} f[\psi(\eta) + i\eta] d\eta,[1]$$

and this equation holds whatever may be the functions ϕ and ψ determining the path of integration. By reducing in it f also to the form of a complex quantity, we are led to only real integrals, and hence we can apply to the former the rules of differentiation holding for real integrals. We therefore obtain

$$\frac{\delta w}{\delta x} = f[x + i\phi(x)] = f(x + iy),$$

$$\frac{\delta w}{\delta y} = if[\psi(y) + iy] = if(x + iy).$$

Hence $\qquad \dfrac{\delta w}{\delta y} = i \dfrac{\delta w}{\delta x};$

consequently (by § 5) w is a function of z. It then follows, from the second of the propositions stated above, that w can also be considered as a function of the lower limit if the upper one be regarded as constant. Since, further (§ 5),

$$\frac{dw}{dz} = \frac{\delta w}{\delta x},$$

it follows also that $\qquad \dfrac{dw}{dz} = f(z).$

On the other hand, the proposition holding for real integrals, that, when $F(z)$ denotes a function of z the derivative of which is $f(z)$,

$$\int_{z_0}^{z} f(z) dz = F(z) - F(z_0),$$

---

[1] This result follows also from the sum (1), by which the integral is defined, if we separate in it the complex quantities into their constituent parts.

cannot without further limitation be applied to complex integrals, because the values of such integrals, as has already been remarked, depend not only upon the upper and lower limits, but also upon the whole series of intermediate values, *i.e.*, upon the path of integration.

17. In order to examine the influence of the path of integration upon the value of the integral, we shall commence with the following considerations. Let

$$z = x + iy$$

be the variable, accordingly x and y the rectangular co-ordinates of the representing point. If we have a *region* of the plane definitely bounded in some way, which can consist of either one or several sheets, and if P and Q be two real functions of x and y which for all points within the region are finite and continuous, *then the surface integral*

$$J = \iint \left(\frac{\delta Q}{\delta x} - \frac{\delta P}{\delta y} \right) dx dy,$$

extended over the whole area of the region, is equal to the linear integral

$$\int (P dx + Q dy),$$

taken round the whole boundary of the region.

We shall not only prove this proposition for the simplest case, when the region consists of only one sheet and is bounded by a simple closed line, but we shall at the same time take into consideration those cases also in which the boundary consists of several separate closed lines, which can either lie entirely outside of one another, or of which one or more can be entirely enclosed by another. Finally, we shall not exclude the case when the region consists of several sheets which are connected with one another along the branch-cuts. Yet we shall then assume that the region does not contain any branch-points at which the functions P and Q become infinite or discontinuous. It is, however, necessary, in order to include all those cases, to determine more definitely the meaning of *boundary-direction*. If we assume, as is customary, that the positive directions of the x- and the y-axis lie so that an observer stationed at the origin and looking in the positive direction of the x-axis has the positive y-axis on his left, then let us so assume the *positive boundary-direction* that one who traces it in this direction shall always have the bounded area of the region on his left. The same can be expressed thus: At each point of the boundary the normal, drawn into the interior of the area, is situated with reference to the positive direction of the boundary just as is the positive y-axis with reference to the positive x-axis. If, for instance, the boundary consist of an external closed line and a circle lying wholly within the same, so that the points within the circle are external to the bounded area of the region, then on the outer line the positive boundary-direction is that of increasing angles, while on the inner circle it is the opposite, as is shown by the arrows in Fig. 21. Now in the linear integral, which we wish to prove to be equal to the given surface integral, the integration must be extended over the whole boundary in the positive direction as just defined.

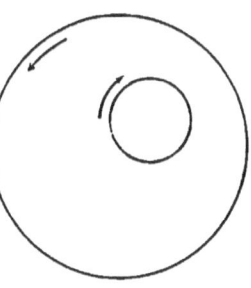

Fig. 21.

We shall write, then, the integral J in the form

$$J = \iint \frac{\delta Q}{\delta x} dx dy - \iint \frac{\delta P}{\delta y} dx dy,$$

and we can then integrate in the first part as to x and in the second part as to y. For this purpose we divide the region into elementary strips, which are formed by straight lines lying infinitely near to one another and, for the first integral, running parallel to the x-axis; in case there are branch-points, we take care to draw such a line through each of them. The whole region is thus divided into infinitely narrow trapezoid-like strips. In Fig. 22, for instance, in a surface consisting of two sheets and bounded by a closed line which makes a circuit twice round a branch-point, several such trapezoid-like pieces are represented, the lines running in the second sheet being dotted. If we now select some one of these elementary strips, belonging to an arbitrary value of y (*i.e.*, in case the surface consists of several sheets, all those elementary strips lying one directly below another in the different sheets which belong to the same value of y), and if we denote the values acquired by the function Q at those places where the elementary strips cut the boundary, counting from left to right (*i.e.*, in the direction of the positive x-axis), at the points of entrance by

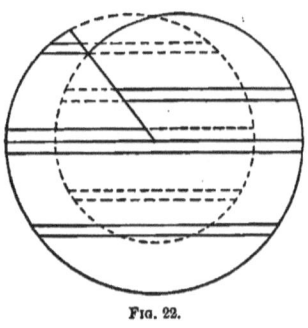

Fig. 22.

$$Q_1, Q_2, Q_3, \ldots,$$

and at the points of exit by

$$Q', Q'', Q''', \ldots,$$

then (Fig. 23)

$$\int \frac{\delta Q}{\delta x} dx = -Q_1 + Q' - Q_2 + Q'' - \cdots;^1$$

therefore,

$$\iint \frac{\delta Q}{\delta x} dx dy = \int -Q_1 dy + \int Q' dy + \int -Q_2 dy + \cdots.$$

[1] It must be noted that this equation remains true, even when $\frac{\delta Q}{\delta x}$ becomes infinite or discontinuous at some place over which the integration extends, if Q suffer no interruption of continuity at this place. If, namely, $f(x)$ be a function of the real variable x, which for $x = a$ is continuous, while its derivative, $f'(x)$, is for the same value discontinuous,

In the integrals on the right y passes through all values from the least to the greatest; therefore dy is always to be taken positively. But if the projections on the y-axis of the elementary arcs which have been cut out from the boundary by the elementary strips be designated in the same sequence as above, at the places of entrance by

$$dy_1, dy_2, dy_3, \cdots,$$

and at the places of exit by

$$dy', dy'', dy''', \cdots,$$

we assume on both sides of a two values x_h and x_k infinitely near to a. If, then, in the integral

$$\int_{x_0}^{x_1} f'(x)dx$$

a lie between the limits x_0 and x_1, and if $f(x)$ remain continuous between the same, while $f'(x)$ is discontinuous only at the place $x = a$, then we can put

$$\int_{x_0}^{x_1} f'(x)dx = \lim \left[\int_{x_0}^{x_h} f'(x)dx + \int_{x_k}^{x_1} f'(x)dx \right],$$

wherein the limit has reference to the coincidence of x_h and x_k with a. Now since $f'(x)$ is continuous from x_0 to x_h and from x_k to x_1, it follows that

$$\int_{x_0}^{x_1} f'(x)dx = \lim \left[f(x_h) - f(x_0) + f(x_1) - f(x_k) \right].$$

Since $f(x)$ is continuous at the place $x = a$, therefore, in passing to the limit, $f(x_h)$ and $f(x_k)$ become equal, or

$$\lim \left[f(x_h) - f(x_k) \right] = 0;$$

therefore, notwithstanding the discontinuity of $f'(x)$ between the limits of the integral, we have

$$\int_{x_0}^{x_1} f'(x)dx = f(x_1) - f(x_0).$$

This case deserves notice here, since it will be shown later that the derivatives of continuous functions can become infinite at branch-points (§ 39).

and if regard be paid to the positive boundary direction (Fig 23), then
$$dy = -dy_1 = -dy_2 = -dy_3 = \cdots$$
$$= +dy' = +dy'' = +dy''' = \cdots;$$
therefore,
$$\iint \frac{\delta Q}{\delta x} dx\, dy = \int Q_1 dy_1 + \int Q' dy' + \int Q_2 dy_2 + \cdots.$$

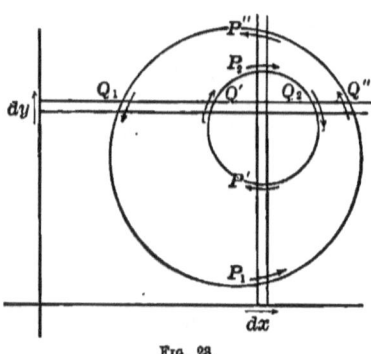

Fig. 23.

In all these integrals y changes in the sense of the positive boundary-direction; therefore they all reduce to a single one, and we have
$$\iint \frac{\delta Q}{\delta x} dx\, dy = \int Q\, dy,$$
if the latter integral be extended along the entire boundary in the positive direction.

In the same manner the second integral
$$\iint \frac{\delta P}{\delta y} dx\, dy$$
can be treated. Here the region is divided into elementary strips by straight lines running parallel to the y-axis, and, as before, such a line is drawn through each branch-point. If, therefore, the values which the function P has at the places

INTEGRALS WITH COMPLEX VARIABLES. 89

where an elementary strip cuts the boundary be designated, in order from below upward (*i.e.*, in the direction of the positive y-axis), at the places of entrance by

$$P_1, P_2, P_3, \cdots,$$

and at the places of exit by

$$P', P'', P''', \cdots,$$

then again

$$\iint \frac{\delta P}{\delta y} dx dy = -\int P_1 dx + \int P' dx - \int P_2 dx + \cdots;$$

and therein dx is positive. But if

$$dx_1, dx_2, dx_3, \cdots, \text{ and } dx', dx'', dx''', \cdots$$

designate the projections of the elementary arcs which are cut out by the elementary strips, then, considering the positive direction of the boundary,

$$dx = + dx_1 = + dx_2 = + dx_3 = \cdots$$
$$= - dx' = - dx'' = - dx''' = \cdots,$$

and therefore

$$\iint \frac{\delta P}{\delta y} dx dy = -\int P_1 dx_1 - \int P' dx' - \int P_2 dx_2 - \cdots,$$
$$= -\int P dx,$$

in which the integral is to be extended in the positive direction round the entire boundary. Combining the two integrals, it follows, as was to be proved, that

$$\iint \left(\frac{\delta Q}{\delta x} - \frac{\delta P}{\delta y} \right) dx dy = \int (P dx + Q dy),$$

the linear integral to be taken round the entire boundary in the positive direction.

This proposition, which is hereby proved for the real functions P and Q, can at once be extended to the case when P and Q are complex functions of the real variables x and y. If we put

$$P = P' + i P'', \quad Q = Q' + i Q'',$$

wherein P', P'', Q', Q'' are real functions of x and y, then

$$\iint \left(\frac{\delta Q}{\delta x} - \frac{\delta P}{\delta y}\right) dx dy = \iint \left(\frac{\delta Q'}{\delta x} - \frac{\delta P'}{\delta y}\right) dx dy \\ + i \iint \left(\frac{\delta Q''}{\delta x} - \frac{\delta P''}{\delta y}\right) dx dy.$$

If the proposition be applied to the right side of the equation, we get

$$= \int (P'dx + Q'dy) + i \int (P''dx + Q''dy) = \int (Pdx + Qdy).$$

We have assumed until now that, within the region under consideration, there are no branch-points or other points at which P and Q are discontinuous. Now, in order to include within our considerations also those regions in which this is the case, it is only necessary to enclose, and thereby exclude, such points by arbitrary small closed lines, these new lines then forming part of the boundary of the region.

18. From the preceding proposition, follows immediately the following: —

(i.) *If $Pdx + Qdy$ be a complete differential, then the integral, $\int (Pdx + Qdy)$, extended over the whole boundary of a region within which P and Q are finite and continuous, is equal to zero.* For, if $Pdx + Qdy$ be a complete differential,

$$\frac{\delta P}{\delta y} = \frac{\delta Q}{\delta x},$$

and therefore all the elements of the surface integral, which is equal to the linear integral, disappear, and accordingly this, as well as that, is equal to zero.

If now $$w = f(z)$$
be a function of a complex variable $z = x + iy$, then [§ 5. (1)]

$$\frac{\delta w}{\delta y} = i \frac{\delta w}{\delta x} = \frac{\delta (iw)}{\delta x},$$

therefore $wdx + iwdy$, i.e., $w(dx + idy)$, or wdz
is a complete differential, and hence

(ii.) $\int f(z)\,dz = 0$, *if this integral be extended round the whole boundary of a region within which $f(z)$ is finite and continuous.*

From this follows further: If the variable z be made to describe between the points a and b two different paths acb and adb (Fig. 24), forming together a closed line which in itself alone is the complete boundary of a region, and if $f(z)$ be finite and continuous within this region, then, for the integral extended round the closed line, we have

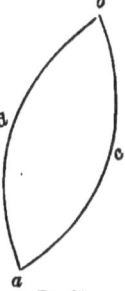

Fig. 24.

$$\int f(z)\,dz = 0.$$

In order to designate briefly an integral taken along a definite path, we shall choose the letter J and add to it the path of integration in parenthesis, so that, for instance, the integral $\int f(z)\,dz$, taken along the path acb, will be designated by $J(acb)$. The last equation can be written

$$J(acbda) = 0.$$

But (§ 16) $J(acbda) = J(acb) + J(bda)$
and $J(bda) = -J(adb)$;
it follows, therefore, that $J(acb) = J(adb)$.

(iii.) *The integral $\int f(z)dz$, therefore, has always the same value along two different paths joining the same points, if the two paths taken together be the boundary of a region in which $f(z)$ is finite and continuous.*

If we have a connected region within which $f(z)$ remains finite and continuous, of such a nature that every closed line described in it forms by itself alone a complete boundary of a part of the region, then the integral $\int f(z)\,dz$ has, along all paths between the two points, the same value. Let the lower limit z_0 be constant; then within such a part of the region the integral is a uniform function of the upper limit, and if $F(z)$ denote a function the derivative of which is $f(z)$, then within this region

$$\int_{z_0}^{z} f(z)\,dz = F(z) - F(z_0),$$

since in this case the value of the integral is independent of the path of integration. The great importance of those surfaces, in which each closed line forms by itself alone the complete boundary of a region, becomes here quite evident.

Riemann has called surfaces of such a character *simply connected surfaces*. Such is, for instance, the surface within a circle. If $f(z)$ be continuous everywhere in such a surface, then, as has been noted,

$$\int_{z_0}^{z} f(z)\,dz$$

is a uniform function of the upper limit. If, on the other hand, $f(z)$ become infinite within the surface of a circle, for instance, only at one point a, and if, in order to obtain a part of the region within which $f(z)$ remains continuous, a small circle k be described round this point, thereby excluding it, then the ring-shaped portion of the plane thus obtained is no longer simply connected; for a line m, which encloses entirely the small circle, does not form by itself alone the entire boundary of a part of the region, but only m and k together. Accordingly the integral extended along m and k together has the value zero; but if the integral extended along k alone be not equal to zero, the integral taken along m cannot be zero. Within such a region, which Riemann has called *multiply connected*,

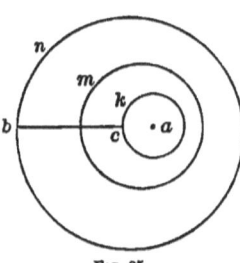

FIG. 25.

the dependence of the integral upon the path of integration continues, and the integral can be regarded as a multiform function of the upper limit.[1]

19. We now drop the assumption that the function $f(z)$ in the region under consideration is everywhere continuous, and we proceed to investigate those integrals of which the paths of integration are boundaries of regions in which the function is not everywhere continuous. If $f(z)$ be infinite or discontinuous at any point of a region, then such a point is to be called a *point of discontinuity*. It may or may not be at the same time a branch-point. If there be points of discontinuity in a region of a plane, we are no longer justified in all cases in concluding that the integral, extended over the whole boundary of the region, has the value zero, because the proof of this proposition rests essentially upon the assumption that $f(z)$ does not become discontinuous within the region. But the following can be proved: —

(iv.) *Whatever may be the value of the integral, it does not change if the region be increased or diminished by arbitrary pieces, provided only that $f(z)$ is finite and continuous within the added or subtracted pieces.* For, if in the first place an added or subtracted piece be completely bounded by one line, as, for instance, A or B (Fig. 26, where $abcda$ is the original boundary), then, if $f(z)$ be continuous within A and B, the integral extended over the boundary of A or B must be zero. The boundary of A or B can therefore be arbitrarily added to the original one without changing the value

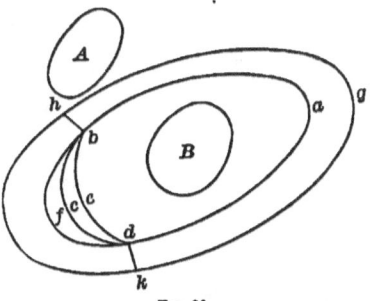

Fig. 26.

---

[1] See Sections IX. and X.

of the integral. If, however, the added or subtracted piece be bounded in part by the original contour, as *bfdcb* or *bcdeb*, then

$$J(bfd) = J(bcd) = J(bed),$$

if $f(z)$ be continuous within these regions. Therefore the portion of the boundary *bcd* can be replaced arbitrarily by either *bfd* or *bed* without altering the value of the integral. From this it follows further that a closed line, which either forms alone the boundary of a region or at least forms part of such a boundary, can also be replaced arbitrarily by a more extended or contracted closed line, provided only that no portions of the surface are thereby either added or subtracted in which $f(z)$ becomes infinite or discontinuous. For, in order to extend, for instance, *abcda* into *ghkg*, it is only necessary to replace first *bcd* by *bhkd*, and then *kdabh* by *kgh*. In a similar manner the validity of the proposition can be proved in all cases. Its general application, however, even to regions which consist of several sheets or contain gaps, can be demonstrated in the following way. If an arbitrary surface T be so divided into two parts, M and N, that $f(z)$ is continuous in M, and if the integral $\int f(z)\,dz$, extended over the boundary of one part, say M, be designated by $J(M)$, then

$$J(M) = 0.$$

If, now, the portions M and N have no common boundary-pieces, the boundaries of M and N together form the boundary of T, and therefore

$$J(T) = J(M) + J(N);$$

consequently also $\quad J(T) = J(N).$

If, however, certain lines C form part of the boundaries of both M and N, then the pieces M and N lie on opposite sides of this line C. If, therefore, the boundaries M and N be described successively in the positive boundary-direction, *i.e.*, so that the bounded region is always on the left, then the lines

C are described twice, in opposite directions; consequently the integrals extended along C cancel each other, while the remaining boundary-pieces of M and N form the entire boundary of T; therefore

$$J(T) = J(M) + J(N),$$

and consequently $\quad J(T) = J(N)$.

Now, just as, according to this, the part M can be separated from the surface T, so, conversely, a surface N can be extended by the addition of a surface M in which the function remains continuous, without changing the boundary-integral.

From this another important proposition can be deduced. If a closed line (I) form by itself alone the complete boundary of a region, and if the function $f(z)$ become discontinuous within it at the points $a_1, a_2, a_3, \cdots$, let each one of these points be enclosed by an arbitrary small closed line, say by a small circle, which, however, in case one of these points of discontinuity be at the same time a branch-point, must be described as many times as there are sheets connected at it; then all these circles, which may be designated by $(A_1), (A_2), (A_3), \cdots$, form, together with the outer line (I), the boundary of a region in which $f(z)$ is continuous (Fig. 27, in which the dotted lines run in the second sheet).

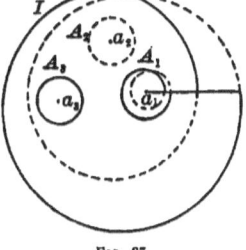

Fig. 27.

Consequently the integral $\int f(z)\,dz$, extended in the positive direction over the whole boundary, is equal to zero. But if the outer line (I) be described in the direction of increasing angles, the small circles $(A_1), (A_2), (A_3), \cdots$ must be described in the direction of decreasing angles. If, therefore, the integral $\int f(z)\,dz$, extended in the direction of the increasing angles

along the lines (I), (A_1), (A_2), (A_3), $\cdots$, be designated by I, A_1, A_2, A_3, $\cdots$, then
$$I - A_1 - A_2 - A_3 - \cdots = 0,$$
and consequently
$$I = A_1 + A_2 + A_3 + \cdots.$$

If now the line (I) be described in a region T, which contains no other points of discontinuity than the above a_1, a_2, a_3, $\cdots$, then the integral I, according to the last proposition, retains its value if it be extended over the boundary of T; we thus obtain the proposition: —

(v.) *The integral $\int f(z) dz$, extended over the whole boundary of a region T, is equal to the sum of the integrals along small closed lines which enclose singly all the points of discontinuity contained within T, all the integrals being taken in the same direction.*

20. By the preceding considerations we are led to the investigation of such closed paths of integration as enclose only one point of discontinuity. We must distinguish, however, whether the point of discontinuity is or is not at the same time a branch-point. Let us consider first a point a, which is not a branch-point, and at which $f(z)$ becomes infinite. If the integral
$$A = \int f(z) dz$$
be taken along a line enclosing one of the points of discontinuity, this line enclosing neither another point of discontinuity nor a branch-point, then the path of integration can be replaced by a small circle described round the point a with the radius r, which can be made to tend towards zero without changing the value of the integral. If we write
$$A = \int (z - a) f(z) \frac{dz}{z - a},$$
and let $\quad z - a = r(\cos\phi + i\sin\phi),$

INTEGRALS WITH COMPLEX VARIABLES.

then r remains constant, and ϕ increases from 0 to 2π when z describes the small circle. Here it is assumed that the point starts from the point z_0, at which the line drawn through a in the positive direction parallel to the principal axis cuts the circle (Fig. 28). This is permissible, since the initial point of the description can be chosen arbitrarily. Now, in order to express $\dfrac{dz}{z-a}$ in terms of z and

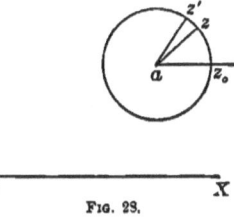

Fig. 29.

ϕ, we remark with Riemann, that dz denotes an infinitely small arc of a circle, starting from any point on the circumference and subtending the angle $d\phi$ at the centre. If the terminal point of this infinitely small arc of the circle be designated by z', then

$$dz = z' - z, \quad \frac{dz}{z-a} = \frac{z'-z}{z-a}.$$

But in § 2, page 21, it has been shown that

$$\frac{z'-z}{z-a} = \frac{\overline{zz'}}{\overline{az}}(-\cos\alpha + i\sin\alpha),$$

wherein α is the angle azz', in this case a right angle; therefore

$$\frac{dz}{z-a} = i\frac{\overline{zz'}}{\overline{az}}.$$

The line $\overline{zz'}$ is an arc of a circle with the angle $d\phi$ at the centre, therefore it equals $rd\phi$, and $\overline{az}$ is equal to the radius r; accordingly we get

$$\frac{dz}{z-a} = i\frac{rd\phi}{r} = id\phi.^1$$

[1] From $\quad z - a = r(\cos\phi + i\sin\phi)$
we also get by direct differentiation, the radius r remaining constant,
$$dz = r(-\sin\phi + i\cos\phi)\,d\phi$$
$$= ir(\cos\phi + i\sin\phi)\,d\phi;$$
therefore $\quad \dfrac{dz}{z-a} = id\phi.$

If this result be substituted in the integral, it follows that

$$A = \int_0^{2\pi} (z-a)f(z)\, id\phi.$$

If the radius r be made to decrease indefinitely, the points z of the circumference of the circle approach the point a; $z-a$ therefore approaches zero, while $f(z)$ becomes infinite. If it happen that $f(z)$ becomes infinite for $z=a$ in such a way that the product $(z-a)f(z)$ tends towards a definite finite limit p, i.e., if

$$\lim\,[(z-a)f(z)]_{z=a} = p,$$

wherein it is expressly assumed that this limiting value always remains the same from whatever side the point z may approach the point a, then we can assume that, for all points z in the vicinity of the point a,

$$(z-a)f(z) = p + \epsilon,$$

wherein ϵ denotes a function of r and ϕ which becomes infinitesimal with r for any value of ϕ. Then

$$A = p \int_0^{2\pi} id\phi + \int_0^{2\pi} \epsilon\, id\phi.$$

If r, and therefore also ϵ, be made to vanish, then the second integral also vanishes, and it follows that

$$A = 2\pi i p.$$

The value of the integral is thereby expressed in terms of the limiting value of $(z-a)f(z)$, when this is finite and determinate. This value of A, by (iv.), does not change if the integration be extended over the complete boundary of a region within which there are no points of discontinuity except a.

The integral
$$\int \frac{dz}{1+z^2}$$
may serve as an example.

Here
$$f(z) = \frac{1}{1+z^2} = \frac{1}{(z-i)(z+i)},$$

INTEGRALS WITH COMPLEX VARIABLES.

which becomes infinite for $z = i$, the point $z = i$ not being a branch-point (the function $\dfrac{1}{1+z^2}$ has no branch-points whatever).

Further,
$$p = \lim\left[\frac{z-i}{1+z^2}\right]_{z=i} = \lim\left[\frac{1}{z+i}\right]_{z=i} = \frac{1}{2i};$$
consequently
$$\int \frac{dz}{1+z^2} = \pi,$$
the integral being extended over a line enclosing the point i in the direction of increasing angles.

If in a region T there be the points of discontinuity $a_1, a_2, a_3, \cdots$, which cannot at the same time be branch-points, and if $f(z)$ become infinite at these points in such a way that the products $(z-a_1)f(z)$, $(z-a_2)f(z)$, $\cdots$ approach determinate finite limiting values $p_1, p_2, \cdots$, i.e., if
$$\lim\,[(z-a_1)f(z)]_{z=a_1} = p_1,$$
$$\lim\,[(z-a_2)f(z)]_{z=a_2} = p_2,$$
$$\cdots\cdots\cdots\cdots,$$
then the integral $\int f(z)\,dz$, extended over the whole boundary of T, assumes the value [(v.), § 19].
$$\int f(z)\,dz = 2\,\pi i\,(p_1 + p_2 + p_3 + \cdots).$$

In the preceding example
$$f(z) = \frac{1}{1+z^2}$$
is infinite also for $z = -i$, and for this point we get
$$p_2 = \lim\left[\frac{z+i}{1+z^2}\right]_{z=-i} = \lim\left[\frac{1}{z-i}\right]_{z=-i} = -\frac{1}{2i};$$
therefore, also,
$$\int \frac{dz}{1+z^2} = -\pi,$$
taken along a line enclosing $-i$.

For a line enclosing both points $+i$ and $-i$ in the direction of increasing angles, this integral becomes $\pi - \pi = 0$.

Now by means of such closed lines as include only a single point of discontinuity it is possible, within a region containing no branch-points nor any gaps, to refer to one another the values of the integrals for the different paths of integration. If two paths bec and bdc (Fig. 29) enclose only one point of discontinuity a and no branch-point,[1] then the one, say bdc, can be replaced, by enclosing the point of discontinuity in a closed line $bghb$ before describing the other path bec. Then by (iv.), § 19,

$$J(bghb) = J(bdceb) = J(bdc) - J(bec),$$

therefore $\quad J(bdc) = J(bghb) + J(bec),$

or also $\quad J(bec) = -J(bghb) + J(bdc)$
$$= J(bhgb) + J(bdc).$$

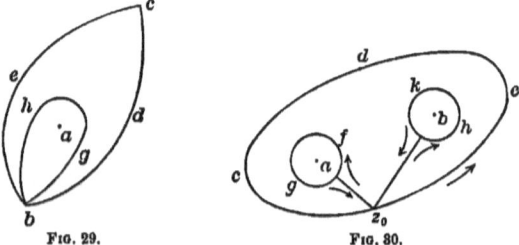

Fig. 29.   Fig. 30.

We get a similar result if two paths enclose several points of discontinuity, but no branch-points. For instance, let the paths $z_0 ed$ and $z_0 cd$ (Fig. 30) enclose two points of discontinuity a and b, and draw from z_0 round each of them a closed line $z_0 fg z_0$ and $z_0 hk z_0$. Then

$$J(z_0 hk z_0) + J(z_0 fg z_0) = J(z_0 edc z_0)$$
$$= J(z_0 ed) - J(z_0 cd);$$

consequently

$$J(z_0 ed) = J(z_0 fg z_0) + J(z_0 hk z_0) + J(z_0 cd).$$

[1] The assumption that the two paths enclose no branch-point is, in general, necessary only in order that together they may form a complete boundary, which may not always be the case if there be branch-points between them.

Therefore the one path can be replaced, by describing closed circuits round each of the points of discontinuity before describing the other path.

The properties of the integral A round a point of discontinuity, if $(z-a)f(z)$ no longer have a determinate finite limiting value at that point, cannot be discussed until later (Section VIII.).

21. We next proceed to the case when the point of discontinuity is at the same time a branch-point, in which case it will be denoted by b, and the value of the integral, for a line described round it, by B. We assume that at this point m sheets of the surface are connected. If we wish to have here a line enclosing the point b, it must make m circuits round b; *e.g.*, let it describe the circumference of a circle m times. Riemann introduces in this case, in the place of z, a new variable ζ, letting

$$\zeta^m = z - b,$$

which therefore receives the value 0 for $z = b$; and he inquires how the function $f(z)$, considered as a function of ζ, behaves at the point $\zeta = 0$. For this purpose we first determine what line is described by ζ when z describes a closed circle, *i.e.*, makes m circuits round ~~the latter.~~

If we let $\quad z - b = r(\cos\theta + i\sin\theta),$

and therefore $\quad \zeta = r^{\frac{1}{m}}\left(\cos\dfrac{1}{m}\theta + i\sin\dfrac{1}{m}\theta\right),$

then r, and consequently also $r^{\frac{1}{m}}$, remains constant, and therefore ζ also describes a circle, namely, one round the origin. But after z has completed one circuit, so that θ has increased from 0 to 2π, then $\dfrac{1}{m}\theta$ has increased from 0 to $\dfrac{2\pi}{m}$; consequently ζ has described the mth part of the circumference. For the second circuit of z, ζ again describes the mth part of the circumference, and likewise for each new circuit of z.

102 THEORY OF FUNCTIONS.

Consequently, after z has made m circuits and returned to its starting-point, ζ has described the entire circumference of the circle exactly once. Therefore, to the m pieces of the region covered by the radius r during these circuits, correspond m sectors of the circle, each with the angle $\dfrac{2\pi}{m}$ at the centre. These join one another and form together a simple circular surface. In Fig. 31, it is assumed that at the point b three sheets are connected, which continue into one another along the branch-cut bb'. The circular lines running in the three sheets have been drawn for the sake of clearness side by side, and the lines running in the 1st, 2d, and 3d sheets are represented by a continuous line, a thickly dotted and a thinly dotted line, respectively. Then

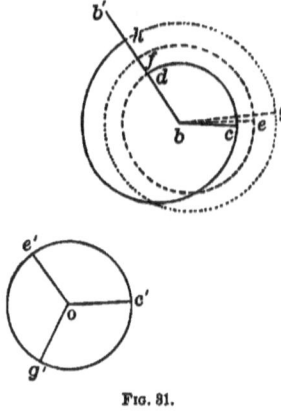

Fig. 31.

to the surface cde corresponds the sector of the circle $c'oe'$,
" " " efg " " " " " " $e'og'$,
" " " ghc " " " " " " $g'oc'$,

and therefore to the whole area of the surface bounded by the closed line $cdefghc$ corresponds the simple circular surface $c'e'g'c'$. It follows therefore that, while z, passing through all the m sheets, returns to its starting-point only after m circuits, ζ does so after the first circuit. The variable ζ therefore does not leave its first sheet, and consequently the function $f(z)$, considered as a function of ζ, does *not* have a branch-point at the place $\zeta=0$. Accordingly, if ζ be introduced as the variable in the integral $\int f(z)\,dz$, extended over a circuit enclosing the branch-point and the point of discontinuity b, the considerations of the last paragraph can be applied, because $\zeta=0$ is not a branch-point, but merely a point of discontinuity. Making the substitution $\zeta = (z-b)^{\frac{1}{m}}$, suppose $f(z)$ changes into $\phi(\zeta)$; then, since $dz = m\zeta^{m-1}d\zeta$,

$$B = m \int \zeta^{m-1}\phi(\zeta)\,d\zeta = m \int \zeta^m \phi(\zeta)\frac{d\zeta}{\zeta}.$$

If, for the sake of brevity, we put $\dfrac{1}{m}\theta = \psi$, and therefore

$$\zeta = r^{\frac{1}{m}}(\cos\psi + i\sin\psi),$$

then ψ increases for the whole circuit from 0 to 2π; and, since as above $\dfrac{d\zeta}{\zeta} = id\psi$, it follows that

$$B = m\int_0^{2\pi} \zeta^m \phi(\zeta)\,id\psi.$$

According to the assumption $\phi(\zeta)$ is infinite for $\zeta = 0$. But if the tendency to become infinite be of such a nature that one of the products

$$\zeta\phi(\zeta),\ \zeta^2\phi(\zeta),\ \cdots,\ \zeta^{m-1}\phi(\zeta)$$

approaches a finite limiting value, then

$$\lim\,[\zeta^m\phi(\zeta)]_{\zeta=0} = 0.$$

Therefore, if the radius r of the circle described round the point b tend towards 0, then $B = 0$.

If we now return to the variable z, we obtain the proposition: *If the integral* $\int f(z)\,dz$ *be extended over a* circuit *enclosing a point of discontinuity, which is at the same time a branch-point at which m sheets of the surface are connected, then the integral has always the value zero, whenever one of the products*

$$(z-b)^{\frac{1}{m}}f(z),\ (z-b)^{\frac{2}{m}}f(z),\ \cdots,\ (z-b)^{\frac{m-1}{m}}f(z)$$

approaches a finite limiting value.

As an example, take

$$\int \frac{dz}{\sqrt{(1-z^2)(1-k^2z^2)}}.$$

Here $$f(z) = \frac{1}{\sqrt{(1-z^2)(1-k^2z^2)}},$$
which becomes infinite for $z = 1$. This point is at the same time a branch-point at which two sheets are connected. If we put
$$\zeta^2 = z - 1,$$
we get $$f(z) = \phi(\zeta) = \frac{1}{i\zeta\sqrt{(2+\zeta^2)(1-k^2[1+\zeta^2]^2)}};$$
so that $\zeta = 0$ is in fact not a branch-point for $\phi(\zeta)$.

Now we get
$$\lim\left[(z-1)^{\frac{1}{2}}f(z)\right]_{z=1} = \lim\left[\frac{1}{i\sqrt{(z+1)(1-k^2z^2)}}\right]_{z=1}$$
$$= \frac{1}{i\sqrt{2(1-k^2)}};$$
therefore $$\lim\left[(z-1)f(z)\right]_{z=1} = 0,$$
and hence also
$$\int \frac{dz}{\sqrt{(1-z^2)(1-k^2z^2)}} = 0,$$
when the integral is taken along a circuit enclosing the point $z = 1$. The integral also acquires the same value when the circuit encloses one of the other branch-points $-1, +\frac{1}{k}, -\frac{1}{k}$.

The investigation of the value of the integral B, in case the conditions of the above proposition are not fulfilled, must be postponed to a later section (Section VIII.).

SECTION V.

THE LOGARITHMIC AND EXPONENTIAL FUNCTIONS.

22. As we shall be obliged to make use of some of the properties of the logarithmic function in the following pages, we must interrupt for a short time the general considerations and

LOGARITHMIC AND EXPONENTIAL FUNCTIONS. 105

take up first the study of this special function. In this connection, it seems to us not unprofitable to make the investigation somewhat more exhaustive than would be necessary for the intended application, and also to add directly to it the consideration of the exponential function, which follows from the logarithmic. Since we shall thus have to deal here with a special case of the general investigations to be taken up in Sections IX. and X., this example may also serve to fix the ideas for those later investigations.

We designate, after Riemann, by the name logarithm a function $f(z)$, which has the property that

$$f(zu) = f(z) + f(u). \tag{1}$$

By this equation the function is entirely determined, except as to a constant, for we shall be able to derive therefrom all its properties. If, in the first place, we let $u = 1$, and leave z arbitrary, it follows that

$$f(z) = f(z) + f(1);$$

therefore $\qquad f(1) = \text{Log } 1 = 0.$

Again, if 0 be substituted for u, we have

$$f(0) = f(z) + f(0);$$

and if we now give z any value for which $f(z)$ is not zero, it will follow that $f(0) = \text{Log } 0 = \infty$; for a similar reason, Log ∞ also becomes infinite. It is further possible to express the logarithm by an integral; for, if equation (1) be differentiated partially as to u, then

$$zf'(zu) = f'(u),$$

and, when $u = 1$,

$$zf'(z) = f'(1).$$

Let us denote the constant $f'(1)$ by m. Upon this constant depends the value of the logarithm of a number. The logarithms of all numbers which can be obtained by assigning to the constant m a definite value form together a system of logarithms, and the constant is called the modulus of the system of logarithms.

From the equation
$$zf'(z) = m$$
follows
$$df(z) = d \operatorname{Log} z = m \frac{dz}{z}; \qquad (2)$$
hence
$$f(z) = m \int \frac{dz}{z} + C.$$

But since $f(1) = 0$, the constant C will become 0, if 1 be taken for the lower limit of the integral, and z be made to assume real values. We write, therefore, in general
$$\operatorname{Log} z = m \int_1^z \frac{dz}{z},$$
and we have thereby expressed the logarithm by a definite integral. For the purposes of analysis, the logarithms of that system are the simplest in which the constant m assumes the value 1. These are called natural logarithms, and will, in what follows, be designated by the term $\log z$. Therefore
$$\log z = \int_1^z \frac{dz}{z},$$
and hence
$$\operatorname{Log} z = m \log z.$$

If we let
$$z = r(\cos \phi + i \sin \phi),$$
we get
$$dz = (\cos \phi + i \sin \phi) dr + r(-\sin \phi + i \cos \phi) d\phi$$
$$= (\cos \phi + i \sin \phi)(dr + ird\phi);$$
hence
$$\frac{dz}{z} = \frac{dr}{r} + id\phi.$$

If z pass along any path from 1 to an arbitrary point z, then r will assume the real values from 1 to r, and ϕ those from 0 to ϕ; therefore
$$\int_1^z \frac{dz}{z} = \int_1^r \frac{dr}{r} + i\phi,$$
or
$$\log z = \log r + i\phi. \qquad (3)$$

By this log z is brought to the form of a complex variable; for, since r assumes only real and positive values in the integral $\int_1^r \frac{dr}{r}$, therefore log r is also real; and it is evident that log r is positive or negative, according as r is greater or less than 1; for, since r is always positive, the representing point moves along the positive principal axis, in the first case in the positive direction, in the second case in the negative; and therefore in the first case all elements $\frac{dr}{r}$ are positive, in the latter all are negative.

We see, further, that the logarithm depends upon the path of integration; for, let ϕ denote the value acquired by the angle, when z moves from 1 to z along a line which does not enclose the origin, and for which the angles increase, then $\phi - 2\pi$ will be the value acquired by this angle when the line moves on the other side of the origin, *i.e.*, in the direction of decreasing angles, from 1 to z; and if a line wind n times round the origin in the direction of increasing angles, then ϕ acquires at z the value $\phi + 2n\pi$. Accordingly

$$\log z = \log r + i\phi \pm 2n\pi i.$$

Our general considerations are thus confirmed. The function $\frac{1}{z}$ has no branch-points, but has the point of discontinuity $z = 0$. If z be made to describe a circuit round the origin, the value of the integral extended over this line in the direction of increasing angles is $2\pi i$, since

$$p = \lim \left[z \cdot \frac{1}{z} \right]_{z=0} = 1. \qquad (\S\ 20)$$

By means of the considerations established at the close of § 20, the same result is obtained as above.

Now from this it follows that the function log z has at no point of the plane a fully determinate value, and that at any two infinitely near points it can, by means of a suitable arrangement of the path of integration, acquire values which differ from one another by a multiple of $2\pi i$. In order to limit as far as possible this indefiniteness, we suppose a line oq (Fig. 32),

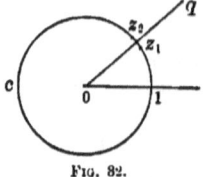

Fig. 32.

which does not cut itself, drawn from the origin and extending to infinity. Such a line is called after Riemann a *cross-cut*. Then, of any two paths leading from 1 to z and enclosing the origin, one must necessarily intersect the cross-cut, and consequently, on all paths not crossing the cross-cut, $\log z$ acquires at each point z a perfectly determinate value, which also changes everywhere continuously with z. But at the points on the cross-cut itself the indefiniteness remains. Now, if the infinite plane in which z moves be designated by T, and be supposed to be actually cut along the cross-cut oq, then a surface arises which may be called T'. In the latter the cross-cut cannot be crossed, and therefore $\log z$ is everywhere a uniform function of z in T', becoming infinite only for $z = 0$ and $z = \infty$, but elsewhere remaining continuous. In the surface T, however, $\log z$ becomes discontinuous on crossing the cross-cut. For, let z_1 and z_2 be two points on the two sides of the cross-cut and infinitely near each other (say z_1 on the right, and z_2 on the left of the direction oq), and let z be made to describe a closed line $1z_1z_2c1$ round the origin in the uncut surface T, starting from 1 and passing through z_1 and z_2; then, according to the above proposition, the integral

$$J(1z_1z_2c1) = 2\pi i,$$

extended along this line. But we have at the same time, since z_1 and z_2 are infinitely near each other,

$$J(1z_1z_2c1) = J(1z_1) + J(z_2c1) = J(1z_1) - J(1cz_2),$$

and consequently $J(1z_1) - J(1cz_2) = 2\pi i$.

If, then, w_1 and w_2 denote the values which $\log z$, now regarded as in T', acquires at z_1 and z_2, so that

$$w_1 = J(1z_1), \quad w_2 = J(1cz_2),$$

we have
$$w_1 - w_2 = 2\pi i.$$

If the surface T be now supposed to be restored, then $\log z$, when z moves from z_1 to z_2, abruptly changes from w_1 into $w_1 - 2\pi i$, or when z moves from z_2 to z_1, abruptly changes from w_2 into $w_2 + 2\pi i$. This holds at whatever place the path of integration may cross the cross-cut. Along the entire cross-cut, therefore, $\log z$ is discontinuous, the values of $\log z$ being greater by $2\pi i$ for all points on the right side than for those on the left. This constant value, by which all values of the function on the one side exceed the neighboring ones on the other side, has been called by Riemann the *modulus of periodicity* of the function, or of the integral, if the former be represented by an integral.

23. The exponential function can be derived from the logarithmic in the following way. By the symbol a^w is to be understood such a function of w that

$$\log (a^w) = w \cdot \log a.$$

Now, if e denote the real number for which $\log e$ has the value 1, and accordingly if e be defined by the equation

$$\int_1^e \frac{dr}{r} = 1,$$

then it follows that $\quad \log (e^w) = w.$

Therefore e^w is the inverse function of the logarithm; for, from $e^w = z$, follows $w = \log z$. From equation (2) (for $m = 1$)

$$\frac{d \log z}{dz} = \frac{dw}{dz} = \frac{1}{z}$$

we get
$$\frac{dz}{dw} = z;$$

consequently
$$\frac{de^w}{dw} = e^w.$$

If we assume for z a complex quantity having the modulus 1, *i.e.*, if we let $\quad z = \cos \phi + i \sin \phi,$

we have, in equation (3), to substitute $r = 1$, and therefore $\log r = 0$. Accordingly,

$$\log(\cos\phi + i\sin\phi) = i\phi,$$

and consequently $\quad \cos\phi + i\sin\phi = e^{i\phi}.$

The exponential function is periodic; for, since to a value of z belongs not only the value w, but also the values $w \pm 2n\pi i$, therefore

$$z = e^w = e^{w \pm 2n\pi i},$$

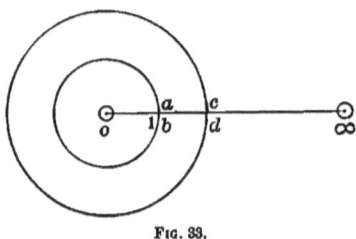

Fig. 33.

and accordingly e^w is not changed if w be increased or diminished by a multiple of the modulus of periodicity $2\pi i$. Let us now try to represent the z-surface T'' on the w-plane W. For this purpose we take as the cross-cut, for greater simplicity, a straight line passing through o and 1 (Fig. 33). If

$$z = r(\cos\phi + i\sin\phi),$$

then $\quad w = \log r + i\phi.$

Consequently $\log r$ and ϕ are the rectangular co-ordinates of a point w. Then, if z be made to describe a circle with radius 1 round the origin in the direction of increasing angles from a to b, $\log r = 0$, and therefore w is a pure imaginary and moves along the y-axis from o to $2\pi i$ (Fig. 34). Again, if z move from a along the left side of the cross-cut to infinity, ϕ remains $= 0$, $\log r$ passes from 0 through the positive values to infinity, and therefore w describes the positive part of the principal axis. But if z move from a along the

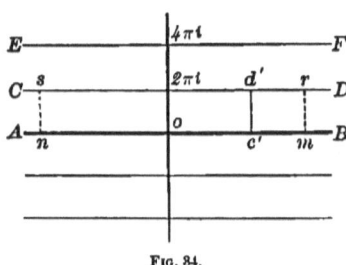

Fig. 34.

left side of the cross-cut to o, then w describes the negative part of the principal axis to infinity. But if z first arrive at b round the origin on the right side of the cross-cut and then pass along its right side to ∞ or o, w first moves on the y-axis from o to $2\pi i$ and then, ϕ constantly remaining equal to 2π, describes a line parallel to the principal axis, first in the positive and then in the negative direction. To the two sides of the cross-cut in T', therefore, correspond in W two different lines, *i.e.*, to the left side the principal axis AB, to the right a straight line CD running through $2\pi i$ parallel to the principal axis (Fig. 34). If z be now made to pass at any place from the left side c of the cross-cut to the right side d by describing a circle round the origin, then r, and therefore also $\log r$, remains constant and ϕ increases from 0 to 2π. Consequently w describes a line $c'd'$ parallel to the y-axis, beginning at the principal axis AB and terminating at the parallel line CD. It follows, therefore, that to all points z in the entire infinite extent of the surface T', in which ϕ cannot increase beyond 2π, correspond only such points w as lie within the strip formed by the two parallel lines AB and CD. The function e^w, or z, thus assumes within this strip all its possible values, and, indeed, each but once, since to any two different values of $w = \log r + i\phi$ belong also different values of r and ϕ, and therefore also different values of

$$e^w = z = r(\cos\phi + i\sin\phi).$$

If we wish to bound the surface T', this can be effected, on the one hand, by describing round the

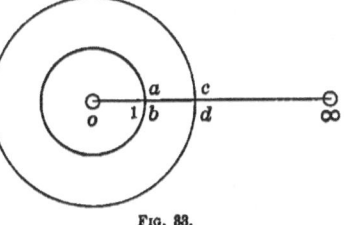

Fig. 33.

origin a circle with a very small radius ρ. To this corresponds in W, since ρ remains constant, a line na running parallel to the y-axis between the two parallel lines AB and CD, and very far removed from the origin on the negative side. This moves to infinity when ρ tends towards zero, *i.e.*, when the circle shrinks into the origin. At all points of this line us, which has been removed to infinity, e^w has therefore the value zero. On the other hand, the boundary of T' can be formed by a circle round the origin with a very large radius R. To this corresponds in w a straight line mr on the positive side, which is very far removed and is parallel to the y-axis. If R increase indefinitely, this straight line also moves to infinity, and at all points on it e^w is infinite. The surface T' can be assumed closed at infinity; then the circle with the large radius R is represented by a small circle round the point ∞, which shrinks into this point when R increases indefinitely. Therefore the two sides of the cross-cut extending from o to ∞ form alone the boundary of a spherical surface T', and to the latter corresponds the strip between the parallel lines AB and CD extending on both sides to infinity.

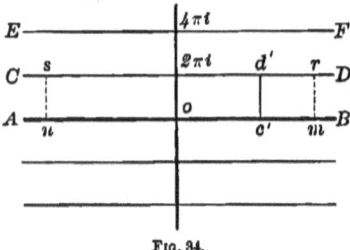

FIG. 34.

If we now increase the angle ϕ beyond 2π, the function w, or $\log z$, proceeds continuously. Then the cross-cut can be supposed to be like a branch-cut, across which the surface T' is continued into another sheet. In this second sheet, then, all relations are the same as in the first, except that at all points in it ϕ is greater by 2π, and accordingly w by $2\pi i$, than at the corresponding places in the first sheet. Therefore we obtain a second strip between the parallel lines CD and EF, which pass through $2\pi i$ and $4\pi i$. By continuing this mode of treatment and applying it also to negative values of ϕ, the plane W is divided into an infinite number of parallel strips. In each of them the function e^w assumes all its values once and has the same values at any two corresponding points of two different strips. On the positive side of each strip e^w tends towards infinity, but on the negative side it approaches zero.

SECTION VI.

GENERAL PROPERTIES OF FUNCTIONS.

24. The basis for the following investigations is found in the exceedingly important proposition proved in § 20: If the integral $\int f(z)\, dz$ be extended over the boundary of a region in which $f(z)$ becomes discontinuous only at a point $z = a$, which is not a branch-point, and in such a manner that $(z-a)f(z)$ approaches, for $z = a$, a definite finite limiting value p, independent of the mode of approach to a, then

$$\int f(z)\, dz = 2\pi i p.$$

Now, if $\phi(z)$ be a function which possesses no branch-points in a region T, and which remains finite and continuous both in the interior and along the boundary of T, and if t denote an arbitrary point in this surface, then the function

$$f(z) = \frac{\phi(z)}{z-t}$$

has in T the properties required in the above proposition. It becomes discontinuous only for $z = t$, and since, like $\phi(z)$, it possesses no branch-points within T, t can never fall on such a point; further, $f(z)$ becomes discontinuous for $z = t$ in such a way that

$$(z-t)f(z) = \phi(z)$$

tends towards a definite finite limiting value, namely $\phi(t)$. Therefore

$$\int \frac{\phi(z)\, dz}{z-t} = 2\pi i \phi(t),$$

and consequently

$$\phi(t) = \frac{1}{2\pi i} \int \frac{\phi(z)\, dz}{z-t}, \tag{1}$$

the integral being extended over the boundary of T.

The validity of this equation is conditioned upon the supposition that the function $\phi(t)$, which is uniform and continuous within T, has a fully determinate, finite value at any point t of this region, the value being always the same, however the variable may approach this point. It may therefore be here noted that, in the case of uniform functions, this condition is not always satisfied[1] at special points, but at such points the function is always at the same time discontinuous. For instance, since the function e^z, for $z = \infty$, becomes zero or infinite, according as the variable z passes to infinity through negative or through positive values (§ 23), therefore the function $e^{\frac{1}{z}}$, for $z = 0$, acquires the value zero or becomes infinite, according as z approaches zero through the real negative or through the real positive values. Likewise the function

$$\frac{c^2}{c - e^{\frac{1}{z}}},$$

in which c denotes a constant, assumes for $z = 0$, in the former case the value c, in the latter case the value zero. At such a point, however, the continuity also always ceases. For, if in the above example the variable be made to increase through the real values, the function, at the passage through the value $z = 0$, is suddenly changed from c into 0.

Thus the requirement that a function be everywhere continuous in a region, at the same time excludes the occurrence of such points.

Now, if the above conditions be fulfilled, equation (1) gives the value of the function ϕ at any point t in the interior of T by an integral, in which the variable z passes through only the points on the boundary of T; this integral has indeed a finite value at every point t situated in the interior of T, and changes continuously with t, as will be proved later. Let the function $\phi(z)$ be given, not by an expression, but by its values for the points of a certain region; then it follows from the

---

[1] This multiplicity of values of the function has nothing in common with that discussed in Section III., which is brought about, in the case of multiform functions, by a multiplicity of paths. By the introduction of Riemann surfaces this kind of multiplicity is removed.

GENERAL PROPERTIES OF FUNCTIONS. 115

above equation that, if the function be given only for all points of the boundary of T, it can also be ascertained for all points in the interior of T, and consequently cannot longer be arbitrarily assumed in the interior of T.

For example, if a function $\phi(z)$ have everywhere along the boundary of T the constant value C, we obtain from (1)

$$\phi(t) = \frac{1}{2\pi i} \int \frac{C dz}{z-t} = \frac{C}{2\pi i} \int \frac{dz}{z-t}.$$

But this integral retains its value if the curve of integration be replaced by a circle described round t. Then we have (§ 20)

$$\int \frac{dz}{z-t} = 2\pi i;$$

and consequently, for every value of t,

$$\phi(t) = C.$$

Therefore, if a function be uniform and continuous everywhere in a region T, and if it have the constant value C along the boundary of T, it is also constantly equal to C everywhere in the interior of T. It follows, further, from (1), by differentiation as to t, that

$$\left.\begin{aligned}
\phi'(t) &= \frac{1}{2\pi i} \int \frac{\phi(z)}{(z-t)^2} dz, \\
\phi''(t) &= \frac{2}{2\pi i} \int \frac{\phi(z)}{(z-t)^3} dz, \\
\phi'''(t) &= \frac{2 \cdot 3}{2\pi i} \int \frac{\phi(z)}{(z-t)^4} dz, \\
&\cdots\cdots\cdots\cdots\cdots \\
\phi^{(n)}(t) &= \frac{2 \cdot 3 \cdots n}{2\pi i} \int \frac{\phi(z)}{(z-t)^{n+1}} dz.
\end{aligned}\right\} \quad (2)$$

All these integrals extend over the boundary of T, while t lies in the interior of T; consequently in them $z-t$ never vanishes. Therefore, if h denote any positive integer, then

$$\frac{1}{(z-t)^h}$$

116   *THEORY OF FUNCTIONS.*

is finite for every value of t considered, and changes continuously with t. The same holds if the above fraction be multiplied by any value $\phi(z)$ which is independent of t; consequently the sum represented by the integral

$$\int \frac{\phi(z)\,dz}{(z-t)^k},$$

in which $\phi(z)$ has to assume in succession all the values occurring along the boundary of T, also changes continuously with t. And since these values are finite according to the assumption, the integral has also a finite value (§ 16). Accordingly all the above integrals, as well as those contained in (1), are finite and continuous functions of t within T^1. From this follows the proposition: *If a function have no branch-points in the interior of a region and be finite and continuous therein, then all its derivatives in the same region are also finite and continuous.*

If in equation (1) the integration be referred to an arbitrarily small circle round the point t with radius r, and if for this purpose we let

$$z - t = r(\cos\theta + i\sin\theta),$$

then

$$\frac{dz}{z-t} = i\,d\theta,$$

and hence

$$\phi(t) = \frac{1}{2\pi}\int_0^{2\pi} \phi(z)\,d\theta.$$

If we now let

$$\phi(z) = u + iv,\quad \phi(t) = u_0 + iv_0,$$

we obtain, on separating the real from the imaginary,

$$u_0 = \frac{1}{2\pi}\int_0^{2\pi} u\,d\theta,\quad v_0 = \frac{1}{2\pi}\int_0^{2\pi} v\,d\theta.$$

Hence it follows that the real components of the function ϕ are, at the point t, the mean values of all the surrounding adjacent values of these components. Therefore u_0 must be

---

[1] C. Newmann, *Vorlesungen über Riemann's Theorie der Abel'schen Integrale*, S. 91.

GENERAL PROPERTIES OF FUNCTIONS. 117

greater than one part and at the same time less than another part of these adjacent values. The same conclusion holds for v_0; and since it also holds at each point of the surface T, the real components of the function ϕ do not have a maximum or a minimum value at any point in T.

25. By means of equation (1) the function ϕ can be developed in a convergent series. Let us describe round an arbitrary point a of the region T a circle, which is still wholly within this region, and therefore does not extend quite to the branch-point or point of discontinuity nearest to a; and let us first take this circle as the curve of integration in equation (1). Now, for every point t lying within the circle,

$$\mod (z - a) > \mod (t - a)$$

(Fig. 35), since z, during the integration, passes through only points on the circumference of the circle; therefore $\overline{az} > \overline{at}$. We can also put

$$\frac{1}{z-t} = \frac{1}{z-a-(t-a)} = \frac{1}{z-a} \cdot \frac{1}{1-\dfrac{t-a}{z-a}},$$

and since

$$\mod \frac{t-a}{z-a} < 1,$$

we can develop this fraction in the convergent series

$$\frac{1}{z-t} = \frac{1}{z-a}\left\{1 + \frac{t-a}{z-a} + \frac{(t-a)^2}{(z-a)^2} + \frac{(t-a)^3}{(z-a)^3} + \cdots\right\}.$$

If this series be substituted in (1), we get

$$\phi(t) = \frac{1}{2\pi i}\left\{\int \frac{\phi(z)\,dz}{z-a} + (t-a)\int \frac{\phi(z)\,dz}{(z-a)^2}\right.$$
$$\left. + (t-a)^2 \int \frac{\phi(z)\,dz}{(z-a)^3} + \cdots\right\}, \quad (3)$$

which is the same as *Taylor's series;* for, according to (1),

$$\frac{1}{2\pi i}\int \frac{\phi(z)\,dz}{z-a} = \phi(a), \qquad (4)$$

and, according to equations (2),

$$\frac{1}{2\pi i}\int\frac{\phi(z)\,dz}{(z-a)^{n+1}}=\frac{\phi^{(n)}(a)}{2\cdot 3\cdots n},\qquad(5)$$

consequently we obtain

$$\phi(t)=\phi(a)+(t-a)\phi'(a)+(t-a)^2\frac{\phi''(a)}{2}+(t-a)^3\frac{\phi'''(a)}{2\cdot 3}+\cdots.\quad(6)$$

This method of deriving Taylor's series has the advantage of showing exactly how far the convergency of the series extends, namely, to all points t which are at a less distance from a than the nearest point of discontinuity or branch-point. In Fig. 35 three such points are marked by crosses. The above-mentioned circle described round a, of which the radius is so chosen that there is no point of discontinuity or branch-point within it or on its circumference, is called the *domain of the point a*.

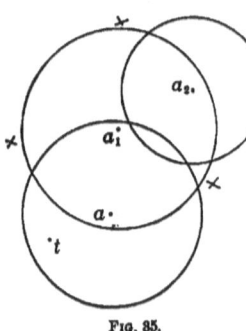

Fig. 35.

The following proposition can then be enunciated: *If a function $\phi(t)$ be finite and continuous at a point a which is not a branch-point, then, for any point t in the domain of a, it can be represented by a convergent series of ascending powers of $t-a$;* for, if we let

$$p_n=\frac{1}{2\pi i}\int\frac{\phi(z)\,dz}{(z-a)^{n+1}},$$

in which the integration is to be extended either along the above circle or along any other line surrounding the point a and enclosing no point of discontinuity or branch-point, then by (3)

$$\phi(t)=p_0+p_1(t-a)+p_2(t-a)^2+\cdots.\qquad(7)$$

GENERAL PROPERTIES OF FUNCTIONS. 119

in which, according to page 116, all the coefficients p have finite values.

Though in the series (3) all integrations must at first be taken along the circle described round a, yet, since the functions

$$\frac{\phi(z)}{z-a}, \frac{\phi(z)}{(z-a)^2}, \frac{\phi(z)}{(z-a)^3}, \text{etc.,}$$

remain finite and continuous up to the point a, the integrals can also be taken along any arbitrarily small circle described round a, without changing their values. It follows that, if the function ϕ be given by its values in an arbitrarily small finite region containing the point a, then all those integrals, and, consequently, all coefficients of the convergent series, are thereby determined, and therefore the value of the function for any point within the large circle can be ascertained.

Now let a_1 be a point which still lies within this circle, then $\phi(t)$ will be known both at a_1 and also in the region immediately contiguous to a_1. Then if a circle be described round a_1, which still leaves outside all points of discontinuity and branch-points (Fig. 35), $\phi(t)$ can be developed in a new series for all points in this circular region. It is evident that by continuing in this way the function $\phi(t)$, which is given only within an arbitrarily small finite part of a region T, can then be determined in the whole region T, when this contains neither a point of discontinuity nor a branch-point.

The same holds if the function $\phi(t)$ be given only along an arbitrarily small finite line proceeding from a. For, if this be the case, let us denote the continuously successive points of this line by a, b, c, d, etc.; then

$$\phi'(a) = \lim \frac{\phi(b) - \phi(a)}{b - a}$$

is therefore known, if $\phi(a)$ and $\phi(b)$ be known. Likewise

$$\phi'(b) = \lim \frac{\phi(c) - \phi(b)}{c - b},$$

by which $\phi'(b)$ is determined. In this manner the values of the derivatives $\phi'(t)$, for all points a, b, c, d, etc., can be found. Then

$$\phi''(a) = \lim \frac{\phi'(b) - \phi'(a)}{b - a},$$

$$\phi''(b) = \lim \frac{\phi'(c) - \phi'(b)}{c - b},$$

etc.,

so that the second derivatives are also known. By continuing in this way we can determine the values of all derivatives for the point a, and consequently of all the coefficients of the series (6). We then obtain, for every point t within the first circle, an expression for $\phi(t)$ in the form of a convergent series. Accordingly we can continue as above and, starting from a small region containing the point a_1, ascertain the value of $\phi(t)$ for all points in the second circle, etc. From the above follows the proposition: *A function of a complex variable, which is given in an arbitrarily small finite portion of the z-plane, can be continued beyond it in only one way.* As a special case of this proposition we emphasize the following: *If a function be constant in a finite arbitrarily small portion of the region T, then it is constant everywhere in T.* For, if it always equal C in a small portion of the surface containing the point a, let us take a circle, described round a and lying within this small region, as the curve of integration in equations (4) and (5), and let

$$z - a = r(\cos\theta + i\sin\theta);$$

then it follows from (4) that

$$\phi(a) = \frac{1}{2\pi}\int_0^{2\pi} \phi(z) d\theta = \frac{C}{2\pi}\int_0^{2\pi} d\theta = C,$$

since $\phi(z)$ possesses the value C at all points on the circumference of the circle. Further, (5) becomes

$$\frac{\phi^{(n)}(a)}{2 \cdot 3 \cdots n} = \frac{1}{2\pi}\int_0^{2\pi} \frac{\phi(z)}{(z-a)^n} d\theta = \frac{1}{2\pi}\frac{C}{r^n}\int_0^{2\pi}(\cos n\theta - i\sin n\theta) d\theta,$$

GENERAL PROPERTIES OF FUNCTIONS.   121

and this value vanishes, since, for every integral value of n different from zero,

$$\int_0^{2\pi} \cos n\theta\, d\theta = 0 \text{ and } \int_0^{2\pi} \sin n\theta\, d\theta = 0.$$

Hence, in the series (3), $\phi(a)$ becomes equal to C, and all other terms disappear; consequently, for any point of the circle of convergence, $\phi(t)$ is equal to C. If the function be continued in the manner indicated above, $\phi(t)$ remains everywhere constantly equal to C. The same holds if $\phi(t)$ be constant along an arbitrarily small finite line. In this case, the above notation being employed, the values $\phi(a)$, $\phi(b)$, $\phi(c)$, etc., are all equal to C, and thus all the derivatives $\phi'(a)$, $\phi''(a)$, etc., again vanish, and thereby also all coefficients of the series (6) except the first, which is equal to C. The same holds therefore as above.

From this special proposition can again be deduced the preceding more general one. For, if two functions $\phi(t)$ and $\psi(t)$ agree in their values in an arbitrarily small portion of a region or of a line, then in this portion the function $\phi(t) - \psi(t)$ is constantly equal to zero; consequently this function is everywhere equal to zero, *i.e.*, $\psi(t)$ is everywhere equal to $\phi(t)$, and therefore the function $\phi(t)$ cannot be continued in two different ways from the portion in which it is given.

26. We now proceed to represent a function, which suffers a *discontinuity of any kind whatever* at a point a (not a branch-point), by a series in the domain of this point.

Let two circles be described round the point a as centre; call the smaller C, the larger K. We assume that the function $\phi(t)$ does not possess a branch-point, either within the smaller circle or in the ring formed by the two circles; further, let $\phi(t)$ be continuous everywhere within the ring, but on the other hand possibly become discontinuous in any way whatever within C. Then the two circles, C and K, bound a region in which $\phi(t)$ satisfies all the conditions under which equation

(1), § 24, holds. We have therefore, at every point t in the interior of the ring,

$$\phi(t) = \frac{1}{2\pi i} \int \frac{\phi(z)\,dz}{z-t},$$

wherein, however, the integral must be extended round each of the circles in the positive boundary-direction, and hence round the small circle in the direction of decreasing angles. Therefore we can put

$$\phi(t) = \frac{1}{2\pi i} \int \frac{\phi(z)\,dz}{z-t} - \frac{1}{2\pi i} \int \frac{\phi(z)\,dz}{z-t} = J_1 + J_2.$$

Then the first integral refers to the circle K, the second to C, and both are to be taken in the direction of increasing angles.

Since, for every point t in the interior of the ring,

$$\mathrm{mod}\,(t-a) < \mathrm{mod}\,(z-a),$$

the first integral J_1 furnishes the same development as was derived in § 25. We thus obtain by (7)

$$J_1 = p_0 + p_1(t-a) + p_2(t-a)^2 + p_3(t-a)^3 + \cdots,$$

wherein
$$p_n = \frac{1}{2\pi i} \int \frac{\phi(z)\,dz}{(z-a)^{n+1}}. \qquad (8)$$

For the second integral J_2, on the other hand, all the points t within the ring lie outside the circle C described by the variable z; hence in this case $\overline{az} < \overline{at}$, or

$$\mathrm{mod}\,(z-a) < \mathrm{mod}\,(t-a) \text{ and } \mathrm{mod}\,\frac{z-a}{t-a} < 1.$$

Therefore, if we put

$$-\frac{1}{z-t} = \frac{1}{t-a-(z-a)} = \frac{1}{t-a} \cdot \frac{1}{1-\dfrac{z-a}{t-a}},$$

we can develop this fraction in a convergent series of ascending powers of $\dfrac{z-a}{t-a}$, and we obtain

$$-\frac{1}{z-t} = \frac{1}{t-a} + \frac{z-a}{(t-a)^2} + \frac{(z-a)^2}{(t-a)^3} + \cdots.$$

GENERAL PROPERTIES OF FUNCTIONS. 123

If we substitute this value in J_2, we get

$$J_2 = \frac{1}{2\pi i}\left\{\frac{1}{t-a}\int \phi(z)\,dz + \frac{1}{(t-a)^2}\int \phi(z)(z-a)\,dz \right.$$
$$\left. + \frac{1}{(t-a)^3}\int \phi(z)(z-a)^2\,dz + \cdots \right\};$$

or if, for the sake of brevity, we let

(9) $$\frac{1}{2\pi i}\int \phi(z)(z-a)^{n-1}\,dz = c^{(n)},$$

$$J_2 = \frac{c'}{t-a} + \frac{c''}{(t-a)^2} + \frac{c'''}{(t-a)^3} + \cdots.$$

Hence we obtain for all points t within the ring the series

(10) $$\phi(t) = p_0 + p_1(t-a) + p_2(t-a)^2 + p_3(t-a)^3 + \cdots$$
$$+ \frac{c'}{t-a} + \frac{c''}{(t-a)^2} + \frac{c'''}{(t-a)^3} + \cdots.$$

This development can be applied when a function $\phi(t)$ suffers a discontinuity of any kind whatever at a point a, which is not a branch-point. For, enclosing the point of discontinuity a in an arbitrarily small circle, the hypotheses previously made are conformed to, if we take this circle as the curve of integration C for the integrals $c^{(n)}$, and refer the integrals p_n to a circle K, which is only so large that every other point of discontinuity occurring (besides a), and every branch-point, lies outside K. Then series (10) furnishes a finite value for $\phi(t)$ at every point t lying within K, with the exception of the point a itself. We remark in this connection that the integrals $c^{(n)}$ can also be taken along the circle K, since they have the same values for it as for the circle C (§ 19).

From the preceding can be derived also a series which holds when $\phi(t)$ suffers any discontinuity at the point $t = \infty$, and when that point is not a branch-point. To this end we let

$$z = \frac{1}{u}, \quad t = \frac{1}{v},$$

whereby $\phi(z)$ changes into $f(u)$, say, and $\phi(t)$ into $f(v)$; then $f(v)$ is discontinuous for $v = 0$. Now let z describe a circle K round the origin, and accordingly let

$$z = r (\cos \theta + i \sin \theta),$$

then
$$u = \frac{1}{r}(\cos \theta - i \sin \theta);$$

hence u likewise describes a circle U round the origin, but in the opposite direction. Since, further, $\frac{1}{r}$ decreases as r increases, to the points z lying outside Z correspond the points u lying within U. Therefore, if we assume the circle Z so large that it encloses all branch-points, and that $\phi(t)$ is discontinuous outside Z only for $t = \infty$, then $f(v)$ has no branch-point within U and suffers a discontinuity only for $v = 0$. Hence[1] we can use series (10) for the development of $f(v)$, if we put $a = 0$, and we obtain

(11) $\quad f(v) = p_0 + p_1 v + p_2 v^2 + p_3 v^3 + \cdots + \dfrac{c'}{v} + \dfrac{c''}{v^2} + \dfrac{c'''}{v^3} + \cdots,$

wherein, by (8) and (9),

$$p_n = \frac{1}{2\pi i} \int \frac{f(u)\,du}{u^{n+1}}, \quad c^{(n)} = \frac{1}{2\pi i} \int f(u)\, u^{n-1} du.$$

Both integrals, according to the remark made above, can be taken round the circle U, which in this case takes the place of the circle K; they are to be taken, like (8) and (9), in the direction of increasing angles. If we introduce z and t again in place of u and v, then

$$du = -\frac{dz}{z^2};$$

therefore $\quad \dfrac{du}{u^{n+1}} = - z^{n-1} dz, \quad u^{n-1} du = - \dfrac{dz}{z^{n+1}}.$

---

[1] We remark that, since $u = 0$ is not a branch-point according to the assumption, we can so draw the branch-cuts that none of them meet the point $u = 0$; then the line U, and therefore also the line Z, bounds a portion of the surface.

The integrals, to be taken as to z, are then extended round the circle Z, but in the direction of decreasing angles, since U was described in the opposite direction. If we wish to take them also in the direction of increasing angles, we have to erase the minus signs, and we then obtain

(12) $\quad p_n = \dfrac{1}{2\pi i} \displaystyle\int \phi(z) z^{n-1} dz, \quad c^{(n)} = \dfrac{1}{2\pi i} \displaystyle\int \dfrac{\phi(z)}{z^{n+1}} dz,$

and hence from (11)

(13) $\quad \phi(t) = p_0 + \dfrac{p_1}{t} + \dfrac{p_2}{t^2} + \dfrac{p_3}{t^3} + \cdots + c't + c''t^2 + c'''t^3 + \cdots.$

This series represents the value of $\phi(t)$ at all points t (except $t = \infty$) which lie outside such a circle Z, described round the origin, that all finite points of discontinuity and all branch-points are situated within the same.

SECTION VII.

INFINITE AND INFINITESIMAL VALUES OF FUNCTIONS.

A. *Functions without branch-points. Uniform functions.*

27. In the closer examination of points of discontinuity, to which we now turn, we shall at first entirely exclude branch-points from our considerations. These therefore, in general, relate to uniform functions, yet it may be expressly stated that they hold also for multiform functions, as long as the discussion refers to only finite parts of the plane in which there are no branch-points.

If we let the variable z approach a point a, a function $\phi(z)$ either does or does not receive the same value for all paths of approach; and, in the former case, the acquired value can be either finite or infinite. Hence there are, for the behavior of a function $\phi(z)$ at a point a, the following possibilities, and only these: —

(1) The function acquires at a for all paths of approach to this point one and the same finite value.

(2) The function becomes infinite at a for all paths of approach.

(3) The function does not acquire at a the same value for all paths of approach, but can for different paths receive different values.[1] (That this can, in fact, occur has been shown already by examples [§ 24].)

In the first case, and only in this, is the function continuous at the point a; in the two other cases it is discontinuous. There are therefore two, and only two, different kinds of discontinuity, and these are also distinguished by special names.

By a *discontinuity of the first kind*, or a *polar discontinuity*,[2] we understand the case when a function $\phi(z)$ becomes infinite at a for every path of approach of the variable to this point. Such a discontinuity is characterized also by the condition that $\dfrac{1}{\phi(z)}$ is absolutely continuous at $z = a$, and that therefore it acquires the value zero for every path of approach to the point a.

A *discontinuity of the second kind*, or a *non-polar discontinuity*, occurs at a point a when, on the contrary, the value acquired by the function at a can be different, according to the path and manner of approach of the variable to the point a. For instance, if a line map can be drawn through a so that the function acquires for the path ma a value different from that for the path pa, then the function springs abruptly from the former value to the latter, when z passes through a on the line

---

[1] We might also think it possible that the function could become infinite of different orders at a for different paths of approach. But, in addition to the fact that this will later be proved to be impossible, such a case cannot be taken into consideration at present, because the conception of infinity of any definite order cannot yet be introduced. The question at present is rather only the alternative, whether, if the function acquire at a the same value for all paths of approach, this value is finite (zero included), or not.

[2] C. Neumann, *Vorlesungen über Riemann's Theorie der Abel'schen Funktionen*, S. 94.

map, and thereby suffers a discontinuity of the second kind. Such a discontinuity occurs in e^z for $z = \infty$, since e^z becomes infinite, zero, or indeterminate, according to the direction in which z moves away to infinity. For, let $z = r(\cos\phi + i\sin\phi)$, then only r becomes infinite, while ϕ indicates the direction in which z moves away to infinity. Then we obtain

$$e^z = e^{r\cos\phi} e^{ir\sin\phi} = e^{r\cos\phi}[\cos(r\sin\phi) + i\sin(r\sin\phi)],$$

wherein the second factor always maintains a finite value. When r becomes infinite, however, the first factor becomes infinite or zero, according as $\cos\phi$ is positive or negative. If, on the other hand, $\cos\phi = 0$, then $r\cos\phi$, and therefore also the first factor, is quite undetermined. In the function $e^{\frac{1}{z}}$ occurs likewise a discontinuity of the second kind for $z = 0$.

An important property, manifesting itself at places of discontinuity of the second kind, results from the following considerations. If a function $\phi(z)$ be absolutely continuous, and hence also not infinite at a point a, the product $(z-a)\phi(z)$ acquires the value zero at a for all paths of approach. We will now show that the converse also holds, namely, that if

$$\lim (z-a)\phi(z) = 0,$$

for all paths of approach to the point a (which, as is always assumed here, is not a branch-point), the function $\phi(z)$ must be continuous at a. For $(z-a)\phi(z)$ is, according to the assumption, continuous at a, and hence can be represented by a series of ascending powers of $z-a$ converging for all points z in the domain of a (§ 25). Let

$$(z-a)\phi(z) = p_0 + p_1(z-a) + p_2(z-a)^2 + p_3(z-a)^3 + \cdots.$$

Therein p_0 denotes the value of $(z-a)\phi(z)$ for $z = a$; and since this is zero according to the hypothesis, it follows that

$$(z-a)\phi(z) = p_1(z-a) + p_2(z-a)^2 + p_3(z-a)^3 + \cdots,$$

from which is obtained

$$\phi(z) = p_1 + p_2(z-a) + p_3(z-a)^2 + \cdots.$$

128   THEORY OF FUNCTIONS.

Accordingly $\phi(z)$ assumes the finite value p_1 for all paths of approach to the point a, and it is therefore continuous at a. We thus obtain the following proposition: *The necessary and sufficient condition to ensure that a uniform function $\phi(z)$ is finite and continuous at a point a is*

$$\lim \left[(z-a)\phi(z)\right]_{z=a} = 0.$$

If we put $(z-a)\phi(z) = F(z)$, we can express this proposition also in the form: If the function $F(z)$ have the value zero at a for all paths of approach to this point, then $\dfrac{F(z)}{z-a}$ is continuous at a; and conversely.

From this now follows: *If a function $\phi(z)$ suffer a non-polar discontinuity at a point $z = a$, it must also become infinite for some manner of approach to a.* For, if $\phi(z)$ were to acquire at a for different paths of approach values not only different but finite, then would

$$\lim \left[(z-a)\phi(z)\right]_{z=a} = 0$$

for all paths of approach, and $\phi(z)$ would not suffer any discontinuity at a. Since now the function always becomes infinite for a discontinuity of the first kind, we can express the preceding proposition also in this way: *A uniform function can be discontinuous, only when at the same time it becomes infinite;* for, in the case of a polar discontinuity this always occurs, and for a non-polar, at least by one way of approach.

But the function must be capable of assuming any arbitrarily assigned value at a point of discontinuity of the second kind a. For, if c be such a value, and if $\phi(z)$ suffer a non-polar discontinuity at a, so do also $\phi(z) - c$ and $\dfrac{1}{\phi(z)-c}$, because these functions likewise acquire different values for different paths of approach to a, when this is true of $\phi(z)$. Now since these functions must also once become infinite, $\phi(z) - c$ must once become zero, and therefore $\phi(z)$ must be equal to c for some one way of approach.

We will make this clear by an example, and in this special case seek to determine also what must be the way of approach to ensure that a function acquires an assigned value. To this end we shall consider the function already instanced (p. 114),

$$\frac{c^2}{c - e^{\frac{1}{z}}},$$

in which c denotes an arbitrary constant. This function has a discontinuity of the second kind at the point $z = 0$. Since it must also become infinite here, $e^{\frac{1}{z}}$ must be capable of assuming the arbitrary value c for $z = 0$. We will inquire when this takes place. Not to disguise the general nature of the process by special circumstances, we will assume c to be complex and let

$$c = h + ik,$$

wherein now h and k denote two arbitrarily assigned real values. Then, if we let

$$z = r(\cos \phi + i \sin \phi),$$

r becomes infinitesimal for every way of approach of z to the origin, while the angle ϕ, made by r with the x-axis, indicates the direction in which we approach the origin. We now obtain

$$e^{\frac{1}{z}} = e^{\frac{1}{r}(\cos \phi - i \sin \phi)} = e^{\frac{\cos \phi}{r}}\left[\cos\left(\frac{\sin \phi}{r}\right) - i \sin\left(\frac{\sin \phi}{r}\right)\right];$$

and if this shall equal $h + ik$, the equations

$$e^{\frac{\cos \phi}{r}} \cos\left(\frac{\sin \phi}{r}\right) = h, \quad e^{\frac{\cos \phi}{r}} \sin\left(\frac{\sin \phi}{r}\right) = -k,$$

or

$$e^{\frac{\cos \phi}{r}} = \sqrt{h^2 + k^2}, \quad \tan\left(\frac{\sin \phi}{r}\right) = -\frac{k}{h}$$

must be satisfied. Now $\frac{\cos \phi}{r}$, for a vanishing value of r, can fail to be infinite, only when ϕ simultaneously approaches the angle $\frac{\pi}{2}$; therefore, if we introduce instead of ϕ the angle $\psi = \frac{\pi}{2} - \phi$, which r makes with the y-axis, and denote by a the real value of $\log \sqrt{h^2 + k^2}$, so that a is arbitrarily assumed just as h and k are, we have instead of the former equations to satisfy the following

(1) $$\frac{\sin\psi}{r}=a,\quad \tan\left(\frac{\cos\psi}{r}\right)=-\frac{k}{h}.$$

But the former will be satisfied at once, by letting ψ and r tend towards zero simultaneously in such a way that

(2) $$\psi=ar$$

always, *i.e.*, by letting the point z approach the origin along the *spiral of Archimedes* which is explicitly determined by the value a, and which is tangent to the y-axis at the origin. .

With this relation existing between ψ and r, $\frac{\cos\psi}{r}$ now becomes infinite as r decreases indefinitely, and therefore the tangent-to-this-curve is capable of assuming every value. But if we denote by α the definite arc contained between $-\frac{\pi}{2}$ and $\frac{\pi}{2}$, the tangent of which has the value $-\frac{k}{h}$, so that the arbitrarily assumed values h and k can be replaced by the equally arbitrary quantities a and α, then also

$$\tan(\alpha+n\pi)=-\frac{k}{h},$$

n denoting a positive integer. The second of equations (1) is satisfied, therefore, if we assume

$$\frac{\cos\psi}{r}=\alpha+n\pi,$$

and make r tend towards zero by increasing n indefinitely. If we substitute $\frac{\psi}{a}$ for r conformably with equation (2), we get

$$\psi=\frac{a\cos\psi}{\alpha+n\pi},$$

for which, since $\cos\psi$ differs from 1 only by an infinitesimal of the second order when ψ and r are infinitesimals of the first order, we can also write

(3) $$\psi=\frac{a}{\alpha+n\pi}.$$

Therefore $e^{\frac{1}{z}}$ acquires the assigned value $c = h + ik$, if the point z approach the origin along the spiral of Archimedes $\psi = ar$ in such a way that the radius vector rotates towards the y-axis *per saltum*, while the angle which it makes with this axis is given by the fraction (3), of which the numerator is constantly equal to a and the denominator increases by π with every spring.

28. We shall now show that a uniform function, which is not a mere constant, must become infinite at some point z, by proving the following proposition: *If a uniform function do not become infinite for some finite or infinite value of the variable, it is a constant.* We can in this case suppose the whole infinite extent of the plane to be the domain of the origin and by (7), § 25, assuming $a = 0$, put

(1) $\qquad \phi(t) = p_0 + p_1 t + p_2 t^2 + p_3 t^3 + \cdots,$

wherein $\qquad p_n = \dfrac{1}{2\pi i} \int \dfrac{\phi(z) dz}{z^{n+1}}.$

We can, moreover, enlarge indefinitely the circle round the origin, to which this integral refers, without changing the value of the integral, since a point of discontinuity nowhere occurs. But if we let

$$z = r(\cos\theta + i\sin\theta),$$

and thus $\qquad \dfrac{dz}{z} = i d\theta,$

we get $\qquad p_n = \dfrac{1}{2\pi} \int_0^{2\pi} \dfrac{\phi(z)}{z^n} d\theta\,;$

and if we let all the values of z along the circumference of the circle become infinite, then p_n vanishes for every value of n, with the exception of $n = 0$, since by the hypothesis $\phi(z)$ remains finite round the circumference of·the infinitely great circle. It therefore follows that

$$p_1 = p_2 = p_3 = \cdots = 0,$$

and series (1) reduces to its first term p_0, so that the function acquires the constant value

$$p_0 = \frac{1}{2\pi} \int_0^{2\pi} \phi(z)\, d\theta,$$

for every value of t.

We can base the proof of this proposition also upon equation (1), § 24, namely,

$$\phi(t) = \frac{1}{2\pi i} \int \frac{\phi(z)\, dz}{z - t}.$$

For, if we take this integral along a circle described round the origin, we can enlarge that indefinitely, because of the assumed properties of the function $\phi(t)$. Accordingly, if we let

$$\frac{dz}{z} = i\, d\theta,$$

we obtain $\quad \phi(t) = \dfrac{1}{2\pi} \displaystyle\int_0^{2\pi} \dfrac{z \phi(z)}{z - t} d\theta = \dfrac{1}{2\pi} \displaystyle\int_0^{2\pi} \dfrac{\phi(z)}{1 - \dfrac{t}{z}} d\theta.$

If now the radius of the circle increase indefinitely, all the values of z in the integral will tend towards infinity; hence $\dfrac{t}{z}$ vanishes, and the integral reduces to the above constant value

$$p_0 = \frac{1}{2\pi} \int_0^{2\pi} \phi(z)\, d\theta$$

independent of t.

From this proposition follows immediately: *If a uniform function be not a constant, it must become infinite for some finite or infinite value of the variable.*

Further follows: *A uniform function must assume the value zero for some value of the variable.* For, if $\phi(z)$ be nowhere equal to zero, $\dfrac{1}{\phi(z)}$ is nowhere infinite; therefore $\dfrac{1}{\phi(z)}$ would be a constant, and hence also $\phi(z)$.

Finally: *A uniform function must be capable of assuming any arbitrary value k at least once.* For, were $\phi(z)$ nowhere equal to k, $\phi(z) - k$ would nowhere be equal to zero; therefore it would be constant, and so too would $\phi(z)$.

It should be emphatically stated that these propositions no longer hold absolutely, if complex values of the variable be excluded. If only real values be considered, the uniform function cos z, for instance, does not become infinite and does not assume every arbitrary value, but only the values between -1 and $+1$. Hence there exists here a certain analogy to algebraic equations, in which also the fundamental proposition, that every algebraic equation must have at least one root, and that every equation of the nth degree has n roots, is not generally valid when only real values are considered.

29. We turn now to the consideration of the cases in which, for the function $\phi(z)$, the product $(z-a)\phi(z)$ no longer vanishes at the point $z = a$. Then $\phi(z)$ by § 27 suffers here a discontinuity. Two possibilities now present themselves: either there is a power of $z-a$ with a positive, integral or fractional exponent μ, for which the product

$$(z-a)^\mu \phi(z)$$

has a determinate finite limit, or there is no such power.

We shall first consider the former case. If this occur, let us denote by n the greatest integer contained in μ, so that

$$n \leq \mu < n+1,$$

where the equality holds when μ itself is an integer. We then have

$$\lim \left[(z-a)^{n+1}\phi(z)\right]_{z=a} = \lim \left[(z-a)^{n+1-\mu}(z-a)^\mu \phi(z)\right]_{z=a} = 0,$$

because $n+1-\mu$ is positive. But if we divide by $z-a$, then, according to § 27, p. 128,

$$(z-a)^n \phi(z)$$

is a function which remains finite for $z = a$. If we denote by $c^{(n)}$ the finite limiting value of the same for $z = a$, then

$$(z-a)^n \phi(z) - c^{(n)}$$

is a function which vanishes for $z = a$, and therefore by § 27

$$(z-a)^{n-1}\phi(z) - \frac{c^{(n)}}{z-a}$$

remains finite for $z = a$. Then, if we denote by $c^{(n-1)}$ the finite limiting value of the same,

$$(z-a)^{n-1}\phi(z) - \frac{c^{(n)}}{z-a} - c^{(n-1)}$$

vanishes for $z = a$, and therefore

$$(z-a)^{n-2}\phi(z) - \frac{c^{(n)}}{(z-a)^2} - \frac{c^{(n-1)}}{z-a}$$

remains finite at the place $z = a$. If we continue in this manner, we finally arrive at a function

$$\phi(z) - \frac{c^{(n)}}{(z-a)^n} - \frac{c^{(n-1)}}{(z-a)^{n-1}} - \frac{c^{(n-2)}}{(z-a)^{n-2}} - \cdots - \frac{c''}{(z-a)^2} - \frac{c'}{z-a},$$

which is finite, and hence also continuous, for $z = a$. Therefore, if we let

$$\phi(z) - \frac{c'}{z-a} - \frac{c''}{(z-a)^2} - \frac{c'''}{(z-a)^3} - \cdots - \frac{c^{(n)}}{(z-a)^n} = \psi(z),$$

$\psi(z)$ denotes a function which is finite and continuous for $z = a$; and if for brevity we let

(1) $\quad \dfrac{c'}{z-a} + \dfrac{c''}{(z-a)^2} + \dfrac{c'''}{(z-a)^3} + \cdots + \dfrac{c^{(n)}}{(z-a)^n} = A,$

we obtain

(2) $\quad \phi(z) = A + \psi(z),$

wherein

$$c^{(n)} = \lim\,[(z-a)^n \phi(z)]_{z=a},$$

$$c^{(n-1)} = \lim\left[(z-a)^{n-1}\phi(z) - \frac{c^{(n)}}{z-a}\right]_{z=a},$$

$$c^{(n-2)} = \lim\left[(z-a)^{n-2}\phi(z) - \frac{c^{(n)}}{(z-a)^2} - \frac{c^{(n-1)}}{z-a}\right]_{z=a},$$

etc.

INF. AND INF'L VALUES OF FUNCTIONS. 135

By this means a part A, which becomes infinite only for $z = a$, is separated from $\phi(z)$, the additional part $\psi(z)$ remaining finite for $z = a$. Now if the finite constant $c^{(n)}$ do not have the value zero, *i.e.*, if the term $\dfrac{c^{(n)}}{(z-a)^n}$ be not wanting in the expression A, we can say: *If* $\lim\ [(z-a)^n\phi(z)]_{z=a}$ *be neither zero nor infinite, the function $\phi(z)$ is infinite of the nth order for $z = a$*. In that case, however, this condition is not satisfied for any fractional exponent μ, but $\lim\ (z-a)^\mu\phi(z)$ is either zero or infinite; for, if $\mu > n$, as we originally assumed, then

$$\lim\ [(z-a)^\mu\phi(z)]_{z=a} = \lim\ [(z-a)^{\mu-n}(z-a)^n\phi(z)]_{z=a} = 0,$$

but if $\mu < n$, then

$$\lim\ [(z-a)^\mu\phi(z)]_{z=a} = \lim \left[\frac{(z-a)^n\phi(z)}{(z-a)^{n-\mu}}\right]_{z=a} = \infty.$$

Therefore $\phi(z)$ cannot be infinite of a fractional order, and the proposition follows: *If a uniform function become infinite of a finite order, it can be infinite only of an integral order.*

An example may be added to the preceding theory. The function

$$\phi(z) = \frac{1}{z^3(z-1)^2}$$

is uniform and has the points of discontinuity $z = 0$ and $z = 1$. For $z = 0$ we have

$$c''' = \lim\ [z^3\phi(z)]_{z=0} = \lim \left[\frac{1}{(z-1)^2}\right]_{z=0} = 1;$$

therefore c''' is finite and not zero, and hence $\phi(z)$ is infinite of the third order for $z = 0$. Now since

$$\lim \left[\frac{1}{(z-1)^2} - 1\right]_{z=0} = 0,$$

we obtain after dividing by z the finite value

$$c'' = \lim \left[\frac{1}{z(z-1)^2} - \frac{1}{z}\right]_{z=0} = 2.$$

In like manner

$$c' = \lim\left[\frac{1}{z^2(z-1)^2} - \frac{1}{z^2} - \frac{2}{z}\right]_{z=0} = 3,$$

and finally

$$\frac{1}{z^3(z-1)^2} - \left(\frac{1}{z} + \frac{2}{z^2} + \frac{3}{z}\right) = \frac{4 - 3z}{(z-1)^2};$$

accordingly the separation into the two parts A and $\psi(z)$ is the following:

$$\phi(z) = \frac{1}{z^3(z-1)^2} = \left(\frac{3}{z} + \frac{2}{z^2} + \frac{1}{z^3}\right) + \frac{4 - 3z}{(z-1)^2}.$$

For the other point of discontinuity, $z = 1$, we have

$$\lim\left[(z-1)^2 \phi(z)\right]_{z=1} = \lim\left(\frac{1}{z^3}\right)_{z=1} = 1,$$

and therefore $\phi(z)$ is infinite of the second order for $z = 1$. After division by $z - 1$ we obtain

$$c' = \lim\left[\frac{1}{z^3(z-1)} - \frac{1}{(z-1)}\right]_{z=1} = \lim\left[-\frac{z^2 + z + 1}{z^3}\right]_{z=1} = -3,$$

and then

$$\frac{1}{z^3(z-1)^2} - \left[\frac{1}{(z-1)^2} - \frac{3}{(z-1)}\right] = \frac{3z^2 + 2z + 1}{z^3}.$$

Therefore the separation in this case is the following:

$$\phi(z) = \frac{1}{z^3(z-1)^2} = \left[-\frac{3}{z-1} + \frac{1}{(z-1)^2}\right] + \frac{3z^2 + 2z + 1}{z^3}.$$

In the cases considered, where $\phi(z)$ becomes infinite of the nth order for $z = a$, *the discontinuity is always a polar;* for if we let

$$(z - a)^n \phi(z) = F(z),$$

$F(z)$ assumes a definite finite value different from zero for all paths of approach to a, and therefore

$$\frac{1}{\phi(z)} = \frac{(z-a)^n}{F(z)}$$

INF. AND INF'L VALUES OF FUNCTIONS. 137

acquires the value zero for all paths of approach. Consequently $\phi(z)$ suffers a discontinuity of the first kind (p. 126). From this it follows further that, when $\phi(z)$ is infinite of the nth order for $z = a$, we can let

$$\phi(z) = \frac{F(z)}{(z-a)^n},$$

wherein $F(z)$, for $z = a$, is finite and not zero, and conversely.

This form, which we can give the function $\phi(z)$ in the case considered, warrants the assumption that an infinity of the nth order can be looked upon as a coincidence of n points, at each of which $\phi(z)$ is infinite of the first order, or as an infinity of multiplicity n. For, if $\phi(z)$ become infinite of the first order at two points a and b, say, we can conformably with the above principles let

$$\phi(z) = \frac{F(z)}{(z-a)},$$

wherein $F(z)$, for $z = a$, is not infinite, but is so for $z = b$, and that of the first order. Therefore we have further,

$$F(z) = \frac{F_1(z)}{z-b}, \quad \phi(z) = \frac{F_1(z)}{(z-a)(z-b)},$$

wherein $F_1(z)$ is not infinite or zero at a or at b. Now if the points a and b coincide, it follows that

$$\phi(z) = \frac{F_1(z)}{(z-a)^2},$$

and therefore $\phi(z)$, by the above criterion, is infinite of the second order at a.

We saw above that, when a function $\phi(z)$ is infinite of a finite order for $z = a$, it suffers here a discontinuity of the first kind; we will now show that the converse is also true. If $\phi(z)$ have a polar discontinuity at the point $z = a$, then $\frac{1}{\phi(z)}$ is continuous, and has the value zero at this place. We can, therefore, by (7), § 25, let

(3) $\quad \dfrac{1}{\phi(z)} = p_1(z-a) + p_2(z-a)^2 + \cdots + p_n(z-a)^n + \cdots;$

for the first term p_0 must be wanting, since it has the value acquired by $\frac{1}{\phi(z)}$ at $z = a$, and this is zero. Of the following coefficients, some may also be zero. Let the first which does not vanish be p_n. Such a coefficient must exist, otherwise $\frac{1}{\phi(z)}$ would be constant, and would have the value zero for every value of z. Therefore let

$$\frac{1}{\phi(z)} = p_n(z-a)^n + p_{n+1}(z-a)^{n+1} + \cdots,$$

wherein p_n is finite and different from zero. Now if we bring this to the form

$$\frac{1}{\phi(z)} = (z-a)^n [p_n + p_{n+1}(z-a) + \cdots]$$

and let
$$\frac{1}{p_n + p_{n+1}(z-a) + \cdots} = F(z),$$

we have
$$\phi(z) = \frac{F(z)}{(z-a)^n};$$

but since $F(z)$, for $z = a$, acquires the value $\frac{1}{p_n}$, finite and different from zero, then $\phi(z)$ becomes by the above criterion infinite of the nth order, and therefore of a finite order.[1]

Consequently the occurrence of a polar discontinuity at a point a is always characterized by the property that the function becomes infinite of a finite order at that point.

From this it follows at once that the case mentioned on p. 126 (note), that $\phi(z)$ always becomes infinite at a point a for different paths of approach to this point, but infinite of different orders, is in fact not possible, but introduces a contradiction. In that case $\frac{1}{\phi(z)}$ would receive the value zero for all paths of approach to a. But, as was shown above, $\phi(z)$ becomes infinite of a definite order determined by that coefficient which is the first in (3) not to vanish.

[1] Königsberger, *Vorlesungen über die Theorie der ell. Funkt.*, I. S. 121.

INF. AND INF'L VALUES OF FUNCTIONS. 139

We now turn our attention to the second possibility mentioned on p. 133, namely, that there is no power of $z-a$ with a finite, positive exponent μ, for which the product $(z-a)^{\mu}\phi(z)$ acquires a finite value for all paths of approach to a. According to the preceding, this can occur only in the case of a discontinuity of the second kind. But the series derived (10), § 26, holds for the latter, because for that development the discontinuity occurring at a could be an entirely arbitrary one, the point a having been excluded by means of a small circle C.

If in (10), § 26, we let

$$p_0 + p_1(z-a) + p_2(z-a)^2 + \cdots = \psi(z),$$

so that $\psi(z)$ represents a finite and continuous function for $z = a$, we obtain

(4) $$\phi(z) = \frac{c'}{z-a} + \frac{c''}{(z-a)^2} + \frac{c'''}{(z-a)^3} + \cdots + \psi(z).$$

In this, by (9), § 26,

$$c^{(n+1)} = \frac{1}{2\pi i} \int \phi(z)(z-a)^n dz,$$

the integral being taken along the circle C described round a. If we substitute in that integral

$$z - a = r(\cos\theta + i\sin\theta), \quad \frac{dz}{z-a} = id\theta,$$

we have $$c^{(n+1)} = \frac{1}{2\pi} \int_0^{2\pi} \phi(z)(z-a)^{n+1} d\theta.$$

Now if, in order in the first place to consider the former case from this point of view, $\phi(z)$ be infinite of the nth order for $z = a$, then $(z-a)^n \phi(z)$ is finite at a, and therefore $(z-a)^{n+1}\phi(z)$ is zero. Hence, if the radius of the circle C be made to tend towards zero, $c^{(n+1)}$ and with greater reason all succeeding coefficients, $c^{(n+2)}$, $c^{(n+3)}$, $\cdots$, vanish. Therefore the series contained in (4) ends with the term $\dfrac{c^{(n)}}{(z-a)^n}$ and changes into the expression A, found previously under (1). If, on the contrary, the second possibility already mentioned occur, in which $(z-a)^n\phi(z)$ does not have a finite limit for any finite value of n, then none of the coefficients $c^{(n)}$ vanish, and the infinite series contained in (4) enters in place of the former expression A. In this case $\phi(z)$ is infinite of an infinitely high order for $z = a$, and at the same time, as remarked above, the discontinuity at a is of the second kind.

Therefore the two kinds of discontinuity are also characterized by this, that in the first occurs an infinity of a finite order, in the second one of an infinitely high order.

We now return to equation (2),

$$\phi(z) = A + \psi(z),$$

in which A denotes either the finite series (1)

$$A = \frac{c'}{z-a} + \frac{c''}{(z-a)^2} + \cdots + \frac{c^{(n)}}{(z-a)^n},$$

or by (4) an infinite series of the same form; $(\psi)z$, however, denoting a finite and continuous function at a. This equation shows that a uniform function $\phi(z)$, which becomes infinite at a place a, is distinguished from a function $\psi(z)$, which remains finite there, only by an expression of the form A. Hence it becomes infinite only as this expression A does. For example, if $\phi(z)$ be infinite of the first order for $z = a$, so that $\lim [(z-a)\phi(z)]_{z=a}$ is finite and not zero, we can then also say that $\phi(z)$ becomes infinite there just as $\dfrac{c'}{z-a}$ does. Or, if $\phi(z)$ be infinite of the second order for $z = a$, it is then infinite either as $\dfrac{c'}{z-a} + \dfrac{c''}{(z-a)^2}$, or only as $\dfrac{c''}{(z-a)^2}$ is. If we have another uniform function $f(z)$, which likewise becomes infinite of the nth order for $z = a$, this can also become infinite only as a similar expression A does, which can differ from the former only in the value of the coefficients c. If the latter function $f(z)$ be given, the coefficients c are thereby also given; therefore $\phi(z)$ is known at a place of discontinuity a, if a function $f(z)$ be given, which becomes infinite at this place just as $\phi(z)$ does. We can then let

$$\phi(z) = f(z) + \psi(z),$$

wherein $\psi(z)$, for $z = a$, remains finite and continuous.

INF. AND INF'L VALUES OF FUNCTIONS.   141

From the equation $\phi(z) = A + \psi(z)$

follows by differentiation

$$\phi'(z) = \frac{dA}{dz} + \psi'(z),$$

where
$$\frac{dA}{dz} = -\frac{c'}{(z-a)^2} - \frac{2c''}{(z-a)^3} - \frac{3c'''}{(z-a)^4} - \cdots - \frac{nc^{(n)}}{(z-a)^{n+1}}.$$

Now since (by § 24) $\psi'(z)$ remains finite for $z = a$, because $\psi(z)$ is here finite, we have: *The derivative $\phi'(z)$ of a uniform function $\phi(z)$ at a place $z = a$, where $\phi(z)$ is infinite, becomes likewise infinite, and that of an order higher by unity than $\phi(z)$.* At all finite points at which $\phi(z)$ is finite, $\phi'(z)$ also remains finite (by § 24), and hence *the finite points of discontinuity of a uniform function are identical with those of its derivative $\phi'(z)$.*

30. We now proceed to the inquiry, how a uniform function $\phi(z)$ behaves for an infinite value of the variable z. We can lead this investigation back to the preceding by putting $z = \frac{1}{u}$, whereby $\phi(z)$ may change into $f(u)$, and then examining $f(u)$ at the point $u = 0$. Now, in the first place, $f(u)$ is finite for $u = 0$ (by § 27) when $\lim [uf(u)]_{u=0} = 0$. Therefore

$$\phi(z) \text{ is finite for } z = \infty \text{ when } \lim \left[\frac{\phi(z)}{z}\right]_{z=\infty} = 0.$$

Further, $f(u)$ becomes infinite of the nth order or of multiplicity n (by § 29) when $\lim [u^n f(u)]_{u=0}$ is neither zero nor infinite.

Hence $\phi(z)$ *is infinite of the nth order for* $z = \infty$ *when*

$$\lim \left[\frac{\phi(z)}{z^n}\right]_{z=\infty}$$

is neither zero nor infinite.

Moreover, we can (by § 29) in this case put

$$f(u) = \frac{Q'}{u} + \frac{Q''}{u^2} + \frac{Q'''}{u^3} + \cdots + \frac{Q^{(n)}}{u^n} + \lambda(u),$$

where $\lambda(u)$ denotes a function which remains finite for $u = 0$, and the quantities Q constant coefficients. If $\lambda(u)$, expressed in terms of z, change into $\psi(z)$, we obtain from the preceding equation the following:

(1) $\quad \phi(z) = Q'(z) + Q''z^2 + Q'''z^3 + \cdots + Q^{(n)}z^n + \psi(z),$

wherein $\psi(z)$ remains finite for $z = \infty$. In this case, therefore, $\phi(z)$ is infinite just as an integral function of z is.

From equation (1) follows

(2) $\quad \phi'(z) = Q' + 2Q''z + 3Q'''z^2 + \cdots + nQ^{(n)}z^{n-1} + \psi'(z).$

To inquire, in the first place, how the derivative $\psi'(z)$ of the function $\psi(z)$ (which remains finite at infinity) behaves at that point, we introduce again the variable u. Since

$$\frac{du}{dz} = -\frac{1}{z^2} = -u^2$$

and $\qquad \psi(z) = \lambda(u),$

we have $\qquad \psi'(z) = -u^2 \lambda'(u).$

Now $\lambda(u)$ is finite for $u = 0$, therefore by § 24 $\lambda'(u)$ is also finite, and consequently

$$\psi'(z) = 0 \text{ for } z = \infty.$$

Therefore, *if a uniform function be finite at the point $z = \infty$, its derivative is equal to zero at that point.* For example,

$$\frac{z^2 + z + 1}{2z^2 - 1}.$$

Then it follows from (2) that $\phi'(z)$ is infinite of an order less by unity than $\phi(z)$, at $z = \infty$. Therefore, if $\phi(z)$ be infinite of the first order only, $\phi'(z)$ remains finite for $z = \infty$.

The integral function of z occurring in (1) can be derived also from the series obtained § 26 (13), which holds when $\phi(z)$ suffers a discontinuity of any kind at $z = \infty$; it is valid then for all points z lying outside a circle which encloses all finite points of discontinuity. If we denote by $\psi(z)$ the first part of that series and put

$$\psi(z) = p_0 + \frac{p_1}{z} + \frac{p_2}{z^2} + \frac{p_3}{z^3} + \cdots,$$

this function remains finite for $z = \infty$ and assumes the value p_0. Denoting the other coefficients by Q instead of by c, we therefore obtain from (13), § 26,

(3) $\qquad \phi(z) = Q'z + Q''z^2 + Q'''z^3 + \cdots + \psi(z),$

wherein by (12), § 26,

$$Q^{(n)} = \frac{1}{2\pi i} \int \frac{\phi(z)}{z^{n+1}} dz,$$

the integral to be taken along a circle round the origin, outside which there is no point of discontinuity except $z = \infty$. By substituting therein $\frac{dz}{z} = id\theta$, we obtain

$$Q^{(n)} = \frac{1}{2\pi} \int_0^{2\pi} \frac{\phi(z)}{z^n} d\theta.$$

Now if $\phi(z)$ be infinite of the nth order for $z = \infty$, then $\lim \left[\frac{\phi(z)}{z^n}\right]_{z=\infty}$ is finite, and therefore $\lim \left[\frac{\phi(z)}{z^{n+1}}\right]_{z=\infty}$ is zero. Consequently, if we let the circle of integration enlarge indefinitely, $Q^{(n+1)}$, $Q^{(n+2)}$, etc., vanish, and the series contained in (3) changes into the integral function in (1).

But when $\phi(z)$ is infinite of an infinitely high order, and when therefore it suffers a discontinuity of the second kind, then in place of the integral function in (1) there enters the series of integral ascending powers of z in (0).

31. From the preceding investigation we now deduce the following propositions: *If a uniform function become infinite* *for no finite value of z, but only for $z = \infty$, and that only of a finite order (multiplicity n), then it is an integral function of the nth degree.* For we have in this case by (1), § 30,

$$\phi(z) = Q'z + Q''z^2 + Q'''z^3 + \cdots + Q^{(n)}z^n + \psi(z).$$

But since $\psi(z)$ is a uniform function, which does not become infinite either for a finite or for an infinite value of z, it is by § 28 a constant. Denoting it by Q, we have

$$\phi(z) = Q + Q'z + Q''z^2 + Q'''z^3 + \cdots + Q^{(n)}z^n;$$

thus $\phi(z)$ is in fact an integral function of the nth degree. Conversely, an integral function of the nth degree becomes infinite only for $z = \infty$, and that of multiplicity n; for

$$\lim \left[\frac{\phi(z)}{z^n} \right]_{z=\infty} = Q^{(n)},$$

and this limit is finite and at the same time different from zero, when $\phi(z)$ is of a degree not less than the nth.

If a uniform function $\phi(z)$ become infinite only for $z = \infty$, but that of an infinitely high order, then it can be developed in a series of powers of z converging for every finite value of z. For in this case the series (3), § 30, holds for all finite values of z, and $\psi(z)$ must be a constant for the same reason as before.

32. *If a uniform function become infinite only for a finite number of values of the variable, and for each only of a finite order (in short, if it become infinite only a finite number of times), then it is a rational function.*

Let $a, b, c, \cdots, k, l, \infty$ be the values of z for which $\phi(z)$ becomes infinite, $\alpha, \beta, \gamma, \cdots, \kappa, \lambda, \mu$, the respective multiplicities of the infinities; then we can in the first place let

$$\phi(z) = Q'z + Q''z^2 + \cdots + Q^{(\mu)}z^\mu + \psi(z),$$

where $\psi(z)$ is not infinite for $z = \infty$, and therefore is infinite only for $z = a, b, c, \cdots, l$; accordingly we have

$$\psi(z) = \frac{c_1'}{z-a} + \frac{c_1''}{(z-a)^2} + \cdots + \frac{c_1^{(\alpha)}}{(z-a)^\alpha} + \psi_1(z),$$

INF. AND INF'L VALUES OF FUNCTIONS.   145

where now $\psi_1(z)$ is infinite only for $z = b, c, \cdots, l$. Therefore we have further

$$\psi_1(z) = \frac{c_2'}{z-b} + \frac{c_2''}{(z-b)^2} + \cdots + \frac{c_2^{(\beta)}}{(z-b)^\beta} + \psi_2(z).$$

If we continue in this way, we arrive at

$$\psi_{n-1}(z) = \frac{c_n'}{z-l} + \frac{c_n''}{(z-l)^2} + \cdots + \frac{c_n^{(\lambda)}}{(z-l)^\lambda} + \psi_n(z),$$

where $\psi_n(z)$ is no longer infinite at all and therefore is a constant. Denoting this by Q, we obtain, when we combine the above expressions,

$$\phi(z) = Q + Q'z + Q''z^2 + \cdots + Q^{(\mu)}z^\mu$$
$$+ \frac{c_1'}{z-a} + \frac{c_1''}{(z-a)^2} + \cdots + \frac{c_1^{(\alpha)}}{(z-a)^\alpha}$$
$$+ \frac{c_2'}{z-b} + \frac{c_2''}{(z-b)^2} + \cdots + \frac{c_2^{(\beta)}}{(z-b)^\beta}$$
$$+ \cdots \cdots \cdots \cdots \cdots$$
$$+ \frac{c_n'}{z-l} + \frac{c_n''}{(z-l)^2} + \cdots + \frac{c_n^{(\lambda)}}{(z-l)^\lambda};$$

hence $\phi(z)$ is in fact a rational function.

Since a rational function can always be brought to the above form, that is, can be separated into an integral function and partial fractions, it follows also, conversely, that a rational function can always become infinite only a finite number of times.

33. *A uniform function $\phi(z)$ is determined, except as to an additive constant, when for each of its points of discontinuity we are given a function which becomes infinite at this point just as $\phi(z)$ does, but which otherwise remains finite and continuous.*

Let a_1, a_2, a_3, etc., be the points of discontinuity of $\phi(z)$, and suppose the value ∞ to be included among them. Further, let $f_1(z), f_2(z), f_3(z)$, etc., be the given functions, which become infinite at the points a_1, a_2, a_3, etc., respectively, but which are elsewhere finite and continuous. Then, if $\phi(z)$ is to become infinite at a_1 just as $f_1(z)$ does, we can let

$$\phi(z) = f_1(z) + \psi(z),$$

where $\psi(z)$ is not infinite for $z = a_1$. Now, since $f_1(z)$ is finite for $z = a_2$, $\psi(z)$ must be infinite at that point, and that just as $\phi(z)$ is. Hence, if $\phi(z)$ is to be infinite at a_2 just as $f_2(z)$ is, we can let

$$\psi(z) = f_2(z) + \psi_1(z),$$

where now $\psi_1(z)$ does not become infinite for a_1 and a_2, but does for a_3, etc. If we continue in this way, we finally arrive at a function ψ which is a constant, since it no longer becomes infinite at any point. Denoting this constant by C, we obtain

$$\phi(z) = f_1(z) + f_2(z) + f_3(z) + \cdots + C.$$

34. We say that a uniform function $\phi(z)$ becomes *infinitesimal or zero of the nth order* for a value of z, when $\dfrac{1}{\phi(z)}$ becomes infinite of the nth order for this value. For this case, by § 29 and § 30,

$$\left.\begin{array}{l} \lim\left[\dfrac{(z-a)^n}{\phi(z)}\right]_{z=a} \\ \lim\left[\dfrac{1}{z^n\phi(z)}\right]_{z=\infty} \end{array}\right\} \text{ is neither zero nor infinite.}$$

Now, since the reciprocal fractions must also have finite limits different from zero, we have as the conditions to ensure that $\phi(z)$ is infinitesimal or zero of the nth order for a finite value $z = a$, and for $z = \infty$:

$$\left.\begin{array}{l} \lim\left[\dfrac{\phi(z)}{(z-a)^n}\right]_{z=a} \\ \lim\left[z^n\phi(z)\right]_{z=\infty} \end{array}\right\} \text{ is neither zero nor infinite.}$$

From these conditions are derived the former ones for the infinite state of $\phi(z)$ by substituting $-n$ for n; hence we can

INF. AND INF'L VALUES OF FUNCTIONS. 147

consider an infinite value as an infinitesimal value of a negative order, or also conversely.

If $\phi(z)$ become zero of the nth order for $z = a$, and if we let

$$\frac{\phi(z)}{(z-a)^n} = F(z),$$

then according to the above $F(z)$ is a function which is finite and not zero for $z = a$. From this it follows that we can let

$$\phi(z) = (z-a)^n F(z),$$

and therefore remove the factor $(z-a)^n$ from $\phi(z)$. If we replace n by $-n$, we obtain again the condition given on p. 137, that, if $\phi(z)$ be infinite of the nth order for $z = a$, we can let

$$\phi(z) = \frac{F(z)}{(z-a)^n},$$

and conversely.

If $\phi(z)$ become infinitesimal of the nth order for $z = \infty$,

then $\qquad z^n \phi(z) = F(z)$

is finite and not zero for $z = \infty$; and this equation holds also for infinite values, if $-n$ be substituted for n. Hence in this case, for infinitesimal values of $\phi(z)$, we can let

$$\phi(z) = \frac{F(z)}{z^n},$$

and for infinite values

$$\phi(z) = z^n F(z),$$

wherein $F(z)$ denotes a function which remains finite and not zero for $z = \infty$.

35. Closely associated with the preceding is the inquiry, how often in a given region a uniform function becomes infinite/infinitesimal of the first order, in which an infinite/infinitesimal value of the nth order is regarded as an infinite/infinitesimal value of the first order of multiplicity n. This number can be expressed by a definite integral.[1] Within a given region T let the uniform function $\phi(z)$ become $\genfrac{}{}{0pt}{}{\text{infinite}}{\text{infinitesimal}}$ at the points $a_1, a_2, a_3,$ etc., of orders $n_1, n_2, n_3,$ etc., respectively, which are to be taken positively for infinitesimal, negatively for infinite values. We will now consider the integral

$$\int d\log \phi(z) \quad \text{or} \quad \int \frac{\phi'(z)}{\phi(z)} dz,$$

taken over the whole boundary of T. The function $\dfrac{\phi'(z)}{\phi(z)}$ becomes infinite at all points at which $\phi(z)$ is zero,[2] and also at all points at which $\phi'(z)$ is infinite.[3] But by § 24, $\phi'(z)$ remains finite at all points at which $\phi(z)$ is finite, and by § 29 becomes infinite at all points at which $\phi(z)$ is infinite; hence the points of discontinuity of $\phi'(z)$ within T are identical with those of $\phi(z)$. Accordingly $\dfrac{\phi'(z)}{\phi(z)}$ becomes infinite at all the points $a_1, a_2, a_3,$ etc., and only at these. Now by § 19 the above integral, taken along the boundary of T, is equal to the sum of the integrals taken round small circles described round the points a. Let A denote one of these integrals corresponding to the point a, and let n denote the order of the $\genfrac{}{}{0pt}{}{\text{infinite}}{\text{infinitesimal}}$ value of $\phi(z)$ at that point. Then by § 34 we have [4]

$$\phi(z) = (z-a)^n \psi(z),$$

where $\psi(z)$, for $z = a$, remains finite and different from zero.

---

[1] This occurs first in Cauchy's writings, *Comptes rendus*, Bd. 40, 1855, I., "Mémoire sur les variations intégrals des fonctions," p. 656.

[2] It is evident from $\phi(z) = (z-a)^n \psi(z)$ (n being positive) that $\phi'(z)$ is finite when $\phi(z)$ is infinitesimal of the first order, and in general that $\phi'(z)$ is infinitesimal of an order lower by unity than $\phi(z)$. [Tr.]

[3] Because $\phi'(z)$ is infinite of an order higher by unity than $\phi(z)$ (p. 141). [Tr.]

[4] In the relation given, n (as always now in the considerations following) is to be taken positively for infinitesimal, negatively for infinite values. [Tr.]

INF. AND INF'L VALUES OF FUNCTIONS. 149

By means of this relation we obtain

$$A = \int d\log \phi(z) = n\int \frac{dz}{z-a} + \int \frac{\psi'(z)}{\psi(z)} dz,$$

the integral to be taken along a small circle described round a. Since $\psi(z)$ is not zero and $\psi'(z)$ not infinite within the circle of integration, $\frac{\psi'(z)}{\psi(z)}$ is continuous, and hence by § 18

$$\int \frac{\psi'(z)}{\psi(z)} dz = 0.$$

Moreover, by § 20, $\quad \int \frac{dz}{z-a} = 2\pi i,$

and therefore $\quad A = 2\pi i n.$

If we sum up these values for all points a, we obtain

$$\int d\log \phi(z) = 2\pi i(n_1 + n_2 + n_3 + \cdots) = 2\pi i \Sigma n,$$

the integral to be taken along the boundary of T.[1]

[1] At this point we have omitted from the text the following: Therein Σn indicates how often $\phi(z)$ becomes infinite / infinitesimal of the first order, if we regard an infinite / infinitesimal value of the nth order as an infinite / infinitesimal value of the first order of multiplicity n. We therefore have the following proposition: *The integral*

$$\int d\log \phi(z)$$

of a uniform function $\phi(z)$, taken along the boundary of a region T, is equal to $2\pi i$ times the number of points within T at which $\phi(z)$ is infinite / infinitesimal of the first order.

This statement of the result is evidently misleading, because Σn is the algebraic sum of the orders of the infinitesimal values of $\phi(z)$ (the infinite values being regarded as infinitesimal values of negative orders, as often stated). Thus if 5 be the number of points at which $\phi(z)$ becomes zero, and 3 the number of points at which it becomes infinite (account being taken in both cases of any multiplicities), then $\Sigma n = 5 - 3 = 2$. The statement is correct, however, if *only* infinitesimal, or *only* infinite values, be included in the area T. [Tr.]

If we let the letter n refer to the infinitesimal values, and denote by $-\nu$ the orders of the infinite values (since these are to be taken negatively), we obtain

(1) $$\int d \log \phi(z) = 2\pi i \,(\Sigma n - \Sigma \nu).$$

If the function $\phi(z)$ remain finite within T, the term $-\Sigma \nu$ drops out of the preceding formula, and it follows that: *The number of points at which a uniform function $\phi(z)$ is zero of the first order within a region T, in which $\phi(z)$ is continuous, is equal to*

$$\frac{1}{2\pi i} \int d \log \phi(z),$$

taken round the boundary of T.

36. If we understand by the points a all *finite points* at which $\phi(z)$ becomes infinite or infinitesimal, it is still important to determine the behavior of $\phi(z)$ for $z = \infty$. Let us assume that $\phi(z)$ is $\genfrac{}{}{0pt}{}{\text{infinite}}{\text{infinitesimal}}$ of the mth order for $z = \infty$, and again let a positive m refer to an infinitesimal value, a negative m to an infinite value. If the boundary of T be assumed to be a circle round the origin, which encloses all the points a, then in the first place according to the preceding

(1) $$\int d \log \phi(z) = 2\pi i \,(\Sigma n - \Sigma \nu),$$

taken round this circle. Now, if a new variable u be introduced in place of z by the relation

$$z = \frac{1}{u},$$

then to every point z corresponds a point u, and to the point $z = \infty$, the point $u = 0$. Further, if we put

$$z = r\,(\cos\theta + i\sin\theta),$$

we get
$$u = \frac{1}{r}(\cos\theta - i\sin\theta).$$

When z describes a closed line Z round the origin, u describes likewise a closed line U round the origin (because thereby θ increases from 0 to 2π), but in the opposite direction. If the radius vector r be made to increase, θ remaining constant, $\dfrac{1}{r}$ decreases, and conversely; therefore to all points z outside Z correspond points u situated within U. If we now introduce u in place of z in the integral

$$\int d\log\phi(z) \text{ or } \int \frac{\phi'(z)}{\phi(z)} dz,$$

and denote by $f(u)$ the function thereby resulting from $\phi(z)$, we obtain

$$\int d\log f(u) \text{ or } \int \frac{f'(u)}{f(u)} du.$$

In the integral as to z, the curve of integration Z is a circle round the origin enclosing all points a; therefore, in the integral as to u, the curve of integration is also a circle round the origin, which, however, is described in the opposite direction. Hence, if we assume for both integrals the integration in the direction of increasing angles, we have

$$\int d\log \phi(z) = -\int d\log f(u),$$

the first integral taken round the circle Z, the second round the circle U. The circle Z encloses all points a; therefore $\phi(z)$ becomes $\dfrac{\text{infinite}}{\text{infinitesimal}}$ *outside* Z only for $z = \infty$, and hence $f(u)$ *within* U is $\dfrac{\text{infinite}}{\text{infinitesimal}}$ only for $u = 0$. For $z = \infty$, $\phi(z)$ was $\dfrac{\text{infinite}}{\text{infinitesimal}}$ of the mth order, so that

$$\lim\left[z^m \phi(z)\right]_{z=\infty} = \lim\left[\frac{f(u)}{u^m}\right]_{u=0}$$

is finite and not zero; accordingly $f(u)$ is also $\dfrac{\text{infinite}}{\text{infinitesimal}}$ of the mth order for $u = 0$. If we let

$$f(u) = u^m \psi(u),$$

$\psi(u)$ denotes a function which is finite and different from zero for $u = 0$, and therefore everywhere within U. Now it follows as above that

$$\int d\log f(u) = m\int \frac{du}{u} + \int \frac{\psi'(u)}{\psi(u)} du,$$

wherein the second integral vanishes, and the first, taken in the direction of increasing angles, is equal to $2\pi i m$. Accordingly we have

$$\int d\log \phi(z) = -\int d\log f(u) = -2\pi i m.$$

If we equate this result to the value of this integral, taken along the same curve, found in (1), we obtain

(2) $\qquad \Sigma n - \Sigma \nu = -m.$

If now $\phi(z)$ be zero for $z = \infty$, then m is positive, and we have

$$m + \Sigma n = \Sigma \nu;$$

but if $\phi(z)$ be infinite for $z = \infty$, then m is negative, and substituting $-\mu$ for it, we obtain

$$\Sigma n = \mu + \Sigma \nu.$$

In both equations, the left side shows how often $\phi(z)$ becomes zero of the first order in the whole infinite extent of the plane, and the right side, how often this function becomes infinite of the first order; from this follows the proposition: *A uniform function in the whole infinite extent of the plane is just as often zero as it is infinite.* Whence we immediately infer: *A uniform function assumes every arbitrary value k just as often as it becomes infinite.*

For $\phi(z) - k$ becomes infinite as often as $\phi(z)$ does; hence $\phi(z) - k$ is zero just as often as $\phi(z)$ is infinite, and therefore $\phi(z)$ is just as often equal to k.

From this follows immediately the fundamental proposition of algebra; for an integral function of the nth degree becomes infinite only for $z = \infty$, and that of multiplicity n; therefore

INF. AND INF'L VALUES OF FUNCTIONS. 153

it must also become n times zero, and hence *an equation of the nth degree must have n roots.*

37. We can now prove again in another form the proposition already proved in § 32, that a uniform function, which becomes infinite only a finite number of times, must be a rational function.

Let a_1, a_2, a_3, etc., be the finite values of z for which a uniform function $\phi(z)$ becomes infinite or infinitesimal, and let n_1, n_2, n_3, etc., respectively denote the orders of the infinite/infinitesimal values, positive for infinitesimal, negative for infinite values. We can then in the first place by § 34 let

$$\phi(z) = (z - a_1)^{n_1} \psi(z),$$

where $\psi(z)$ is finite and not zero for $z = a_1$, but becomes infinite/infinitesimal for $z = a_2, a_3,$ etc. Then

$$\psi(z) = (z - a_2)^{n_2} \psi_1(z),$$

where now
$$\psi_1(z) = \frac{\phi(z)}{(z-a_1)^{n_1}(z-a_2)^{n_2}}$$

does not become infinite/infinitesimal for a_1 and a_2, but does become infinite/infinitesimal for a_3, etc.; if we continue in this way, we arrive at a function

$$\lambda(z) = \frac{\phi(z)}{(z-a_1)^{n_1}(z-a_2)^{n_2}(z-a_3)^{n_3}\cdots} = \frac{\phi(z)}{\Pi(z-a)^n},$$

which no longer becomes infinite/infinitesimal for any finite value of z. From this can now be shown, however, that it cannot become infinite/infinitesimal for $z = \infty$. For, since

$$(z-a)^n = z^n \left(1 - \frac{a}{z}\right)^n,$$

we can write $\quad \Pi(z-a)^n = z^{\Sigma n} \Pi \left(1 - \frac{a}{z}\right)^n.$

But if m denote the order of the $\genfrac{}{}{0pt}{}{\text{infinite}}{\text{infinitesimal}}$ value of $\phi(z)$ for $z = \infty$, positive for infinitesimal, negative for infinite values, then by (2), § 36,

$$\Sigma n = -m,$$

since here Σn denotes the same number that was there designated by $\Sigma n - \Sigma \nu$. Accordingly we have

$$\Pi(z-a)^n = z^{-m}\Pi\left(1-\frac{a}{z}\right)^n$$

and

$$\lambda(z) = \frac{z^m \phi(z)}{\Pi\left(1-\dfrac{a}{z}\right)^n}.$$

But for $z = \infty$,

$$\lim \frac{z^m \phi(z)}{\Pi\left(1-\dfrac{a}{z}\right)^n} = \lim z^m \phi(z),$$

and this is finite and not zero by § 34, since $\phi(z)$ is $\genfrac{}{}{0pt}{}{\text{infinite}}{\text{infinitesimal}}$ of the mth order for $z = \infty$. Therefore $\lambda(z)$ is in fact a function which remains finite for $z = \infty$; now since it also does not become infinite for any finite value of z, it must by § 28 be a constant. If we denote it by C, we have

$$\phi(z) = C\Pi(z-a)^n.$$

If we retain now a_1, a_2, a_3, etc., for the finite values of z, for which $\phi(z)$ becomes zero of orders n_1, n_2, n_3, etc., respectively; and if we denote by $\alpha_1, \alpha_2, \alpha_3$, etc., the finite values, for which $\phi(z)$ becomes infinite of orders ν_1, ν_2, ν_3, etc., respectively, then we have

$$\phi(z) = C\frac{(z-a_1)^{n_1}(z-a_2)^{n_2}(z-a_3)^{n_3}\cdots}{(z-\alpha_1)^{\nu_1}(z-\alpha_2)^{\nu_2}(z-\alpha_3)^{\nu_3}\cdots}.$$

Therefore $\phi(z)$ is actually a rational function, and appears here with numerator and denominator separated into factors, while in § 32 it was resolved into partial fractions and an integral function.

From this follows, further: *A uniform function is determined except as to a constant factor, when once we know all finite values for which it becomes infinite and infinitesimal, and also the orders of the $\genfrac{}{}{0pt}{}{infinite}{infinitesimal}$ values;* and: *Two uniform functions, which agree in these values and in their orders, are equal to each other except as to a constant factor.*

B. Functions with branch-points. Algebraic functions.

38. An algebraic function w of z is defined by an algebraic equation, in which the coefficients of the powers of w are rational functions of z; therefore by an equation of the form

(1) $\quad w^p - f_1(z)w^{p-1} + f_2(z)w^{p-2} - \cdots + (-1)^p f_p(z) = 0,$

wherein $f_1(z)$, $f_2(z)$, $\cdots$, represent rational functions of z, and in which the coefficient of the highest power of w is assumed to be unity. If $w_1, w_2, \cdots, w_p$ denote the p roots of this equation for any assumed value of z, they are then the p values of the function for the value of z in question.

Of these values at least one must become infinite for some finite or infinite value of z. For we have

$$w_1 + w_2 + \cdots + w_p = f_1(z).$$

Now, since $f_1(z)$, as a uniform function, must by § 28 become infinite for some value of z, so for this value of z at least one of the summands $w_1, w_2, \cdots, w_p$ must become infinite.

But we can show further that w can become infinite only for such a value of z as leads at the same time to an infinite value of at least one of the rational functions $f_1(z)$, $f_2(z)$, $\cdots$. For we have

$$w_1 + w_2 + \cdots + w_p = f_1(z)$$
(2) $\quad w_1 w_2 + w_1 w_3 + \cdots + w_{p-1} w_p = f_2(z)$
$$\cdots \cdots \cdots \cdots$$
$$w_1 w_2 \cdots w_p = f_p(z).$$

If we now denote by w_p one of those values of w which become infinite for a certain value of z, we can remove this and introduce the sums of the combinations of the remaining values of w.

Letting
$$w_1 + w_2 + \cdots + w_{p-1} = \phi_1(z)$$
$$w_1w_2 + w_1w_3 + \cdots + w_{p-2}w_{p-1} = \phi_2(z)$$
$$\cdots \cdots \cdots \cdots$$
$$w_1w_2 \cdots w_{p-1} = \phi_{p-1}(z);$$

we then easily obtain

$$f_1(z) = \phi_1(z) + w_p$$
$$f_2(z) = \phi_2(z) + w_p\phi_1(z)$$
$$\cdots \cdots \cdots \cdots$$
$$f_{p-1}(z) = \phi_{p-1}(z) + w_p\phi_{p-2}(z)$$
$$f_p(z) = w_p\phi_{p-1}(z).$$

Now since w_p is infinite, therefore, by reason of the last equation, either $f_p(z)$ must be infinite, and then the above proposition is already proved; or, if this be not the case, $\phi_{p-1}(z)$ must vanish. In like manner it follows from the next to the last equation, that either $f_{p-1}(z)$ must be infinite or $\phi_{p-2}(z)$ must be zero. Continuing in this way up to the second equation, if the case occur that none of the f-functions from $f_p(z)$ to $f_2(z)$ is infinite for the value of z in question, the ϕ-functions must all vanish from $\phi_{p-1}(z)$ to $\phi_1(z)$, and therefore it follows from the first equation that $f_1(z)$ must be infinite.

Now since the converse also follows from equations (2), viz., that, whenever one of the f-functions becomes infinite, so does at least one of the w-functions, we obtain all the z-points at which the algebraic function w becomes infinite, by looking for all the z-points at which the rational functions $f_1(z), f_2(z), \cdots$ become infinite. But since the latter can become infinite at only a finite number of points, therefore also *an algebraic function can become infinite at only a finite number of points.*[1]

[1] Königsberger, *Vorlesungen über die Theorie der elliptischen Funktionen*, I. S. 112.

It can now be proved also that an algebraic function cannot become infinite of an infinitely high order at any point. For, since the rational f-functions become infinite of only a finite order (§ 32), let a be a point at which occurs the infinity of highest order for these functions, and let this highest order be the $(r-1)$th. Then, for those f-functions which become infinite of this highest order, the product

$$(z - a)^{r-1} f(z)$$

is neither zero nor infinite at $z = a$ (§ 29). For those functions, on the other hand, which are either not infinite at all or infinite of a lower order at $z = a$, this product is zero; and this value holds for all the f-functions, when in place of $r-1$ a higher exponent occurs. Now, if we introduce in equation (1) another function W in place of w, by letting

$$(3) \qquad w = \frac{W}{(z-a)^r},$$

we obtain for W the following equation:

$$W^p - (z-a)^r f_1(z) W^{p-1} + (z-a)^{2r} f_2(z) W^{p-2} - \cdots$$
$$+ (-1)^p (z-a)^{pr} f_p(z) = 0.$$

But since in this the exponent of $z - a$ for every coefficient is greater than $r - 1$, these coefficients all vanish for $z = a$, and the equation reduces to

$$W^p = 0$$

for this value of z, so that the values of W corresponding to $z = a$ are all zero. Now, from (3) follows

$$W = (z - a)^r w;$$

therefore the values of the w-functions have the property that for them the product $(z - a)^r w$ vanishes at the point $z = a$. But since one or more of them are here infinite, there must be for them a positive, integral or fractional exponent μ, *less than r*, for which the product $(z - a)^\mu w$ receives a finite value different from zero; and in such a case we again say, the w-functions involved are infinite of a finite order.[1]

If a denote a point at which none of the f-functions acquires an infinite value of an order as high as $r-1$, but at which, nevertheless, one or more of them become infinite, the statement holds so much the more. But if an infinite value occur for $z = \infty$, the substitution $z = \dfrac{1}{u}$ is made, and then the f-functions become rational functions of u; therefore the previous reasoning is applicable to the point $u = 0$ if the w-functions be also treated as functions of u. Hence we obtain the proposition: *An algebraic function always becomes infinite at a finite number of points, and at each of them infinite of a finite order.*

In the Riemann surface, in which by § 12 an algebraic function can be regarded as a uniform function of position in the surface, we no longer need to examine those points which are not branch-points, since for them the principles of the preceding section which refer only to the finite parts of the plane containing no branch-points do not lose their validity.

Hence we have in this place to examine in detail only the branch-points themselves, and we begin with the investigation made in § 21, which proved the following proposition: if $z = b$ be a branch-point of a function $f(z)$, at which m sheets of the z-surface are connected [a *winding-point* of the $(m-1)$th order (§ 13)], and if we let

$$(z-b)^{\frac{1}{m}} = \zeta,$$

by which $f(z)$ changes into $\phi(\zeta)$, say, then $\phi(\zeta)$ does *not* have a branch-point at the place $\zeta = 0$ corresponding to $z = b$.

Now we can, in the first place, apply to the point $\zeta = 0$ the criterion of p. 128 for the finiteness of the function, and infer that $\phi(\zeta)$ remains finite at the place $\zeta = 0$ if

$$\lim\,[\zeta\phi(\zeta)]_{\zeta=0} = 0;$$

therefore we obtain as the necessary and sufficient condition that $f(z)$ remain finite at the branch-point $z = b$:

$$\lim\,[(z-b)^{\frac{1}{m}}f(z)]_{z=b} = 0.$$

[1] Königsberger, *Vorlesungen*, u. s. w., I. S. 177.

Further, the considerations of § 29 show that if $\phi(\zeta)$ become infinite of the nth order at the point $\zeta = 0$, we can let

$$\phi(\zeta) = \frac{g'}{\zeta} + \frac{g''}{\zeta^2} + \frac{g'''}{\zeta^3} + \cdots + \frac{g^{(n)}}{\zeta^n} + \lambda(\zeta),$$

wherein $\lambda(\zeta)$ remains finite for $\zeta = 0$, and the quantities g denote constant coefficients. Therefore we have

$$f(z) = \frac{g'}{(z-b)^{\frac{1}{m}}} + \frac{g''}{(z-b)^{\frac{2}{m}}} + \cdots + \frac{g^{(n)}}{(z-b)^{\frac{n}{m}}} + \psi(z),$$

wherein $\psi(z)$ equals $\lambda(\zeta)$, say, and remains finite for $z = b$. Then

$\lim (z-b)^{\frac{n}{m}} f(z)$ *is finite and not zero, and the order of the infinity of* $f(z)$ *is denoted by the fraction* $\frac{n}{m}$.

At b, m sheets of the z-surface are connected, hence in this place m function-values become equal. If these be designated by $w_1, w_2, \cdots, w_m$, the quantities

$$w_1(z-b)^{\frac{n}{m}}, \; w_2(z-b)^{\frac{n}{m}}, \; \cdots, \; w_m(z-b)^{\frac{n}{m}}$$

have each a finite limit different from zero, and therefore the same is also true of the product

$$w_1 w_2 \cdots w_m (z-b)^n.$$

Hence we can also say: *The function w becomes infinite of multiplicity n at b where m sheets are connected, if each of the values becoming equal at this place be infinite of the order* $\frac{n}{m}$.

The principle proved on p. 158 can then be expressed as follows: *An algebraic function always becomes infinite a finite number of times.*

We determine more explicitly the kind of infinity of $f(z)$, by specifying the expression by which $f(z)$ differs at b from a function which remains finite at that point. This expression proceeds, as the last equation shows, according to powers of $(z-b)^{-\frac{1}{m}}$. Thus, we say, for instance, that $f(z)$ becomes infinite as $\dfrac{g'}{(z-b)^{\frac{1}{m}}}$, or as $\dfrac{g''}{(z-b)^{\frac{2}{m}}}$, or as $\dfrac{g'}{(z-b)^{\frac{1}{m}}}+\dfrac{g''}{(z-b)^{\frac{2}{m}}}$, etc.

Let us now consider the value $z=\infty$, which, as we have already seen, § 14, can be represented by a definite point, and which can also occur as a branch-point. Let us put

$$z = \frac{1}{u} \text{ and } f(z) = \phi(u);$$

then $u=0$ is a branch-point of the $(m-1)$th order for $\phi(u)$, if $z=\infty$ be such for $f(z)$. Therefore $f(z)$ is finite for $z=\infty$, if

$$\lim\left[u^{\frac{1}{m}}\phi(u)\right]_{u=0} = \lim\left[\frac{f(z)}{z^{\frac{1}{m}}}\right]_{z=\infty} = 0.$$

But, if $f(z)$ for $z=\infty$, and hence also $\phi(u)$ for $u=0$ become infinite of the order $\dfrac{n}{m}$, then

$$\lim\left[u^{\frac{n}{m}}\phi(u)\right]_{u=0} = \lim\left[\frac{f(z)}{z^{\frac{n}{m}}}\right]_{z=\infty}$$

is finite and not zero, and we can let

$$\phi(u) = \frac{g'}{u^{\frac{1}{m}}}+\frac{g''}{u^{\frac{2}{m}}}+\ldots+\frac{g^{(n)}}{u^{\frac{n}{m}}}+\lambda(u);$$

or $\quad f(z) = g'z^{\frac{1}{m}}+g''z^{\frac{2}{m}}+\ldots+g^{(n)}z^{\frac{n}{m}}+\psi(z),\quad (4)$

wherein $\psi(z) = \lambda(u)$ remains finite for $z=\infty$. In this case we say that $f(z)$ becomes infinite of multiplicity n at $z=\infty$.

39. We must now also study the behavior at a branch-point of the derivative $\dfrac{dw}{dz}$, in which w is written for $f(z)$. First let us consider only those finite points at which w remains finite. It has been proved (§ 24) that if w be finite, continuous and

INF. AND INF'L VALUES OF FUNCTIONS. 161

uniform in a region, *i.e.*, if it possess neither points of discontinuity nor branch-points, then $\dfrac{dw}{dz}$ likewise remains finite and continuous in the same region. Now, if we express the derivative by the limiting value to which it is equal, denoting by w_a the value of w corresponding to $z = a$, we have

$$\frac{dw}{dz} = \lim \left[\frac{w - w_a}{z - a}\right]_{z = a}$$

and can accordingly say that when $z = a$ is neither a point of discontinuity nor a branch-point of w, then

$$\lim \left[\frac{w - w_a}{z - a}\right]_{z = a}$$

is not infinite.

But we can also determine under what condition this limiting value is not zero. To that end we have only to consider z as a function of w. If the point $w = w_a$, corresponding to the point $z = a$, be not a branch-point of the function z, then according to the above

$$\lim \left[\frac{z - a}{w - w_a}\right]_{z = a}$$

is not infinite, and hence the reciprocal fraction

$$\lim \left[\frac{w - w_a}{z - a}\right]_{z = a}$$

is not zero.

Hence, we obtain in the first place the following proposition: *If $z = a$ and $w = w_a$ be two finite points corresponding to each other, and if neither $z = a$ be a branch-point of w, nor $w = w_a$ a branch-point of z, then*

$$\lim \left[\frac{w - w_a}{z - a}\right]_{z = a}$$

is finite and not zero.

It follows that the derivative $\dfrac{dw}{dz}$ at a finite point (at which w also is finite) can become zero or infinite, only when at that point a branching occurs, either for w considered as a function of z, or for z as a function of w.

Now, if in place of a a branch-point b enter, at which, however, w has a finite value w_b, let

$$(z-b)^{\frac{1}{m}} = \zeta,$$

the z-surface winding m times round b (by § 21); then w, considered as a function of ζ, has neither a point of discontinuity nor a branch-point at the place $\zeta = 0$. If we assume now the case in which ζ, also regarded as a function of w, does not have a branch-point at the place $w = w_b$, the hypotheses of the preceding proposition are satisfied, and therefore

$$\lim\left[\frac{w-w_b}{\zeta}\right]_{\zeta \doteq 0} = \lim\left[\frac{w-w_b}{(z-b)^{\frac{1}{m}}}\right]_{z \doteq b}$$

is neither zero nor infinite. But now

$$(z-b) = \zeta^m;$$

therefore z is a rational function of ζ and consequently by § 15 just such a branched function of w as ζ is. Therefore, if ζ do not possess a branch-point at the place $w = w_b$, then z, considered as a function of w, likewise has no such point there, and hence the proposition follows:—

(i.) *If w have a winding-point of the $(m-1)$th order at $z = b$, but z, regarded as a function of w, no branch-point at $w = w_b$, then*

$$\lim\left[\frac{w-w_b}{(z-b)^{\frac{1}{m}}}\right]_{z \doteq b} \text{ is finite and not zero.}$$

If this finite and limiting value be denoted by k, then also

$$\lim\left[\frac{(w-w_b)^m}{z-b}\right]_{z \doteq b} = k^m;$$

but $\qquad \lim\left(\frac{w-w_b}{z-b}\right)_{z \doteq b} = \frac{dw}{dz},$

hence $\quad \dfrac{dw}{dz} = \lim\left[\dfrac{k^m}{(w-w_b)^{m-1}}\right]_{w \doteq w_b} = \lim\left[\dfrac{k}{(z-b)^{\frac{m-1}{m}}}\right]_{z \doteq b}.$

INF. AND INF'L VALUES OF FUNCTIONS. 163

Therefore:—

(ii.) *With the hypothesis of proposition* (i.), $\dfrac{dw}{dz}$ *becomes infinite at b, and in such a manner that*

$$\lim\left[(w-w_b)^{m-1}\dfrac{dw}{dz}\right]_{w=w_b} \text{ and } \lim\left[(z-b)^{\frac{m-1}{m}}\dfrac{dw}{dz}\right]_{z=b},$$

are neither zero nor infinite.

Ex. $w = \sqrt[3]{z}$, $\dfrac{dw}{dz} = \dfrac{1}{3\sqrt[3]{z^2}}$, for $z = 0$.

If, on the other hand, ζ or $(z-b)^{\frac{1}{m}}$ possess a branch-point at $w = w_b$, such that μ sheets of the w-surface are connected in it, the hypotheses of proposition (i.) are satisfied, because ζ as a function of w has a winding-point of the $(\mu - 1)$th order at w_b, but w as a function of ζ has no branch-point at $\zeta = 0$;

hence
$$\lim\left[\dfrac{\zeta}{(w-w_b)^{\frac{1}{\mu}}}\right]_{w=w_b},$$

and therefore also the reciprocal fraction

$$\lim\left[\dfrac{(w-w_b)^{\frac{1}{\mu}}}{(z-b)^{\frac{1}{m}}}\right]_{z=b},$$

is finite and not zero. Since now z and ζ are like-branched functions of w, we conclude:—

(iii.) *If w have a winding-point of the $(m-1)$th order at $z = b$, and z as a function of w, a winding-point of the $(\mu - 1)$th order at the corresponding place $w = w_b$, then*

$$\lim\left[\dfrac{(w-w_b)^{\frac{1}{\mu}}}{(z-b)^{\frac{1}{m}}}\right]_{z=b} \text{ is finite and not zero.}$$

If we denote this finite limit by h, then

$$\lim\left[\dfrac{(w-w_b)^{\frac{m}{\mu}}}{z-b}\right]_{z=b} = h^m;$$

and since
$$\lim\left[\frac{w-w_b}{z-b}\right]_{z=b} = \frac{dw}{dz},$$

also $\dfrac{dw}{dz} = \lim\left[h^m(w-w_b)^{\frac{\mu-m}{\mu}}\right]_{w=w_b} = \lim\left[h^\mu(z-b)^{\frac{\mu-m}{m}}\right]_{z=b}.$

Therefore: —

(iv.) *With the hypothesis of proposition* (iii.), $\dfrac{dw}{dz}$ *is zero or infinite, according as* $\mu >$ *or* $< m$, *and in such a manner that*

$$\lim\left[(w-w_b)^{\frac{m-\mu}{\mu}}\frac{dw}{dz}\right]_{w=w_b} \text{ and } \lim\left[(z-b)^{\frac{m-\mu}{m}}\frac{dw}{dz}\right]_{z=b}$$

are neither zero nor infinite.

Ex. $w^3 = z^2$, for $z=0$; $m=3$, $\mu=2$,
$$\frac{dw}{dz} = \frac{2}{3}\frac{1}{w^{\frac{1}{2}}} = \frac{2}{3}\frac{1}{z^{\frac{1}{3}}}.$$

We have still to examine the value $z = \infty$, retaining the hypothesis that it represents a branch-point at which w is finite. Let
$$z = \frac{1}{u},$$
and let w' be the value of w corresponding to $z = \infty$ or $u = 0$. If we assume that $z = \infty$ is a winding-point of the $(m-1)$th order for w, but that $w = w'$ is a winding-point of the $(\mu - 1)$th order for z, we obtain by (iii.), since z and u are like-branched functions of w, and also since the branching of w remains the same: —

(v.) $\quad \lim\left[\dfrac{(w-w')^{\frac{1}{\mu}}}{u^{\frac{1}{m}}}\right]_{u=0}$ or $\lim\left[z^{\frac{1}{m}}(w-w')^{\frac{1}{\mu}}\right]_{z=\infty}$

is finite and not zero.

If this limit be denoted by h, we have by (iv.)
$$\frac{dw}{du} = \lim\left[h^m(w-w')^{\frac{\mu-m}{\mu}}\right]_{w=w'} = \lim\left[h^\mu u^{\frac{\mu-m}{m}}\right]_{u=0}.$$

But now
$$\frac{dw}{dz} = -\frac{1}{z^2}\frac{dw}{du};$$
therefore
$$\frac{dw}{dz} = -\lim\left[\frac{h^m(w-w')^{\frac{\mu-m}{\mu}}}{z^2}\right]_{z=\infty} = -\lim\left[\frac{h^\mu u^{\frac{\mu-m}{m}}}{z^2}\right]_{z=\infty},$$

or
$$\frac{dw}{dz} = -\lim\left[\frac{(w-w')^{\frac{\mu+m}{\mu}}}{h^m}\right]_{w=w'} = -\lim\left[\frac{h^\mu}{z^{\frac{\mu+m}{m}}}\right]_{z=\infty}.$$

Consequently: —

(vi.) *With the hypothesis of* (v.), $\dfrac{dw}{dz}$ *is zero, and in such a manner that the expressions*

$$z^2(w-w')^{\frac{m-\mu}{\mu}}\frac{dw}{dz},\quad \frac{1}{(w-w')^{\frac{\mu+m}{\mu}}}\frac{dw}{dz},\quad z^{\frac{\mu+m}{m}}\frac{dw}{dz}$$

have limits finite and different from zero.

Ex. $(w-w')^2 = \dfrac{1}{z^3}$; $m=3, \mu=2$;
$$\frac{dw}{dz} = -\frac{3}{2}\frac{1}{z^{\frac{5}{2}}} = -\frac{3}{2}(w-w')^{\frac{5}{3}}.$$

Finally, let us turn to the consideration of the case when w itself becomes infinite at a branch-point, and at first let us assume the latter to be finite and equal to b. Now, if m sheets are connected at $z = b$ and μ sheets at $w = \infty$, we can determine directly from (v.) what expression remains finite and different from zero. For, if $z - b$ be there put in place of $w - w'$, and w in place of z, and if further m and μ interchange, it follows that
$$\lim\left[w^{\frac{1}{\mu}}(z-b)^{\frac{1}{m}}\right]_{z=b} = h$$

remains finite and not zero. Now since from this results
$$\lim\left[w(z-b)^{\frac{\mu}{m}}\right]_{z=b} = h^\mu,$$

so that this limit is also neither zero nor infinite, it follows (by § 38) that w is infinite of the order $\frac{\mu}{m}$ in this case; and the converse at the same time holds. Making the same substitutions as above in the second of the expressions (vi.), we see that

$$\frac{1}{(z-b)^{\frac{m+\mu}{m}}}\frac{dz}{dw},$$

and hence also the reciprocal value

$$(z-b)^{\frac{m+\mu}{m}}\frac{dw}{dz},$$

is finite and not zero in the limit, and that therefore $\frac{dw}{dz}$ is infinite of the order $\frac{m+\mu}{m}$. Hence the conclusion is: *If w become infinite of the order $\frac{\mu}{m}$ at a winding-point $z=b$ of the $(m-1)$th order, then the point $w=\infty$ itself is at the same time a branch-point of the $(\mu-1)$th order, and conversely; and $\frac{dw}{dz}$ becomes infinite of the order $\frac{m+\mu}{m}$.*

Secondly, if w become infinite for $z=\infty$, and if this point be a winding-point of the $(m-1)$th order, while $w=\infty$ is a winding-point of the $(\mu-1)$th order, let $z=\frac{1}{u}$; then by the preceding proposition

and hence

$$\lim\left[w\cdot u^{\frac{\mu}{m}}\right]_{u=0}$$

$$\lim\left[\frac{w}{z^{\frac{\mu}{m}}}\right]_{z=\infty}$$

is finite and not zero, and therefore w is infinite of the order $\frac{\mu}{m}$. Further,

$$\lim\left[u^{\frac{m+\mu}{m}}\frac{dw}{du}\right]_{u=0} \text{ and } \lim\left[\frac{1}{z^{\frac{m+\mu}{m}}}\frac{dw}{du}\right]_{z=\infty}.$$

remain finite and not zero.

Now since
$$\frac{dw}{du} = -z^2 \frac{dw}{dz},$$

therefore
$$\lim \left[z^{\frac{m-\mu}{m}} \frac{dw}{dz} \right]_{z=\infty}$$

is finite and different from zero, and hence $\dfrac{dw}{dz}$ is either zero or infinite, according as $m >$ or $< \mu$. *E.g.*, take the equation
$$(w - w')^3 (z - b)^5 = 1;$$
then
$$w - w' = \frac{1}{(z-b)^{\frac{5}{3}}}, \quad z - b = \frac{1}{(w-w')^{\frac{3}{5}}},$$

and therefore w is infinite of the order $\frac{5}{3}$, for $z = b$. At the same time three sheets of the z-surface are connected at the place $z = b$, and at the corresponding place $w = \infty$ five sheets of the w-surface. Further
$$\frac{dw}{dz} = -\frac{5}{3} \cdot \frac{1}{(z-b)^{\frac{8}{3}}},$$

and therefore the derivative is infinite of the order $\frac{8}{3}$ for $z = b$.
For the equations
$$w = z^{\frac{3}{5}}, \text{ and } w = z^{\frac{5}{3}},$$

the places $z = \infty$ and $w = \infty$ correspond. We obtain respectively
$$\frac{dw}{dz} = \frac{3}{5} \frac{1}{z^{\frac{2}{5}}}, \text{ and } \frac{dw}{dz} = \frac{5}{3} z^{\frac{2}{3}};$$

hence, for $z = \infty$, $\dfrac{dw}{dz}$ is zero in the first case, and infinite in the second.

40. We can now specify in what way the z-surface is represented on the w-surface in the vicinity of a branch-point, and thus dispose of the exceptional case mentioned in § 7.

If we assume that $z = b$ is a winding-point of the $(m-1)$th order, and $w = w_b$ a winding-point of the $(\mu-1)$th order, we have, by (3), § 39, for

$$\frac{(w - w_b)^{\frac{1}{\mu}}}{(z - b)^{\frac{1}{m}}}$$

a definite, finite limiting value different from zero. Therefore if z' and z'' be two points lying infinitely near to b in different directions, w' and w'' the points of the w-surface corresponding to them, then it is the above expression, and no longer, as in § 7, the expression $\dfrac{w - w_b}{z - b}$, which has the same finite limit for both pairs of corresponding points. Therefore

$$\frac{(w' - w_b)^{\frac{1}{\mu}}}{(z' - b)^{\frac{1}{m}}} = \frac{(w'' - w_b)^{\frac{1}{\mu}}}{(z'' - b)^{\frac{1}{m}}}$$

or
$$\left(\frac{w' - w_b}{w'' - w_b}\right)^{\frac{1}{\mu}} = \left(\frac{z' - b}{z'' - b}\right)^{\frac{1}{m}}.$$

If we now put $w' - w_b = \rho'(\cos\psi' + i\sin\psi')$,

$$w'' - w_b = \rho''(\cos\psi'' + i\sin\psi''),$$
$$z' - b = r'(\cos\phi' + i\sin\phi'),$$
$$z'' - b = r''(\cos\phi'' + i\sin\phi''),$$

we have
$$\left(\frac{\rho'}{\rho''}\right)^{\frac{1}{\mu}}\left(\cos\frac{\psi' - \psi''}{\mu} + i\sin\frac{\psi' - \psi''}{\mu}\right)$$
$$= \left(\frac{r'}{r''}\right)^{\frac{1}{m}}\left(\cos\frac{\phi' - \phi''}{m} + i\sin\frac{\phi' - \phi''}{m}\right),$$

and therefrom
$$\left(\frac{\rho'}{\rho''}\right)^{\frac{1}{\mu}} = \left(\frac{r'}{r''}\right)^{\frac{1}{m}}, \quad \frac{\psi' - \psi''}{\mu} = \frac{\phi' - \phi''}{m},$$

or
$$\left(\frac{\rho'}{\rho''}\right)^{m} = \left(\frac{r'}{r''}\right)^{\mu}, \quad m(\psi' - \psi'') = \mu(\phi' - \phi'').$$

Hence there no longer exists similarity in the infinitesimal elements in the neighborhood of the branch-point.

In the example cited in § 7,

$$w = z^2,$$

$m = 1$ and $\mu = 2$; therefore $\dfrac{dw}{dz} = 0$, for the branch-point $w = 0$ (corresponding to $z = 0$), since $\mu > m$. At the same time we have

$$\frac{\rho'}{\rho''} = \left(\frac{r'}{r''}\right)^2,\ \psi' - \psi'' = 2\,(\phi' - \phi''),$$

a particular case (§ 7).

An immediate consequence of this is (among others) the proposition:[1] *The angle under which two confocal parabolas intersect is half as large as the angle between their axes.* By the method given in § 7, or also easily in another way, we satisfy ourselves that to each straight line in z which does not pass through the origin, corresponds a parabola in w, the focus of which is at the origin, and the axis of which corresponds to a straight line in z passing through the origin, and at the same time parallel to the former. The angle which two straight lines in z, not passing through the origin, make with each other is just as large as the angle under which the corresponding parabolas intersect; under the same angle also intersect the straight lines in z, passing through the origin, which correspond to the axes of the parabolas in w. But since the origin is a branch-point of z, and in fact $m = 1$ and $\mu = 2$, the axes of the parabolas make with each other an angle twice as large.

41. It was shown above (§ 38) that a multiform algebraic function becomes infinite a finite number of times. We now prove the converse, namely:

If a function w have n values for each value of z, and become infinite only a finite number of times, it is an algebraic function.

[1] Siebeck: "Ueber die graphische Darstellung imaginärer Funktionen," *Crelle's Journ.*, Bd. 55, S. 230.

Let us denote by $w_1, w_2, w_3, \cdots, w_n$ the n values of w corresponding to a definite value of z. If we form the product
$$S = (\sigma - w_1)(\sigma - w_2) \cdots (\sigma - w_n),$$
wherein σ denotes an arbitrary quantity independent of z, then S is symmetric with regard to $w_1, w_2, \cdots, w_n$. Now let z describe any apparently closed line (§ 12), then some or all of the values $w_1, w_2, \cdots, w_n$ will have changed, but at the n points of the z-surface situated one immediately above another, w will again have the same values, but in a different sequence; consequently S, regarded as a function of z, has not changed. S is therefore one-valued at all points, and hence is a uniform function of z. In addition, S becomes infinite only when one or more of the functions $w_1, w_2, \cdots, w_n$ become infinite. Each of the latter, according to the assumption, becomes infinite only a finite number of times; hence the same is true also of S. Therefore S is a uniform function which becomes infinite only a finite number of times; and hence by § 32 it is a *rational function* of z. If now $z = a$ be a point of discontinuity of w which is not at the same time a branch-point, and if w_a be infinite of multiplicity a at this point, then

$$w_a(z-a)^a, \text{ and therefore also } (\sigma - w_a)(z-a)^a,$$

is not infinite at a (§ 29). If, further, $z = b$ be at the same time a point of discontinuity and a branch-point, and if μ sheets are connected in it, then μ values of w also become equal. If these be denoted by $w_1, w_2, \cdots, w_\mu$, and if the number of times that w becomes infinite at b be denoted by β, then by § 38 the quantities

$$w_1(z-b)^{\frac{\beta}{\mu}}, w_2(z-b)^{\frac{\beta}{\mu}}, \cdots, w_\mu(z-b)^{\frac{\beta}{\mu}},$$

and therefore also

$$(\sigma - w_1)(z-b)^{\frac{\beta}{\mu}}, (\sigma - w_2)(z-b)^{\frac{\beta}{\mu}}, \cdots, (\sigma - w_\mu)(z-b)^{\frac{\beta}{\mu}},$$

are not infinite. Consequently the product

$$(\sigma - w_1)(\sigma - w_2) \cdots (\sigma - w_\mu)(z-b)^\beta$$

also remains finite for $z = b$.

INF. AND INF'L VALUES OF FUNCTIONS. 171

Now let $\quad a_1, a_2, \ldots, a_\lambda$

denote the points of discontinuity which are not branch-points, and

$$b_1, b_2, \ldots, b_\nu$$

the points of discontinuity which are at the same time branch-points; further, let the respective orders α and β be designated by corresponding subscripts. Then if we multiply S by the product

$$Z = (z - a_1)^{\alpha_1}(z - a_2)^{\alpha_2} \cdots (z - a_\lambda)^{\alpha_\lambda}$$
$$\times (z - b_1)^{\beta_1}(z - b_2)^{\beta_2} \cdots (z - b_\nu)^{\beta_\nu},$$

the product

$$SZ = (\sigma - w_1)(\sigma - w_2) \cdots (\sigma - w_n)$$
$$\times (z - a_1)^{\alpha_1}(z - a_2)^{\alpha_2} \cdots (z - a_\lambda)^{\alpha_\lambda}$$
$$\times (z - b_1)^{\beta_1}(z - b_2)^{\beta_2} \cdots (z - b_\nu)^{\beta_\nu}$$

remains finite for all values a and b, and therefore for all finite values of z. Consequently SZ is a uniform function which becomes infinite only for $z = \infty$, and that of a finite order; therefore SZ is (by § 31) an *integral function* of z. Now in the first place in SZ each factor of Z becomes infinite for $z = \infty$; if h denote the number of times that w becomes infinite for $z = \infty$, then the number of times that SZ becomes infinite for $z = \infty$ is

$$h + \Sigma\alpha + \Sigma\beta,$$

and this number is exactly the number of times that w becomes infinite altogether. For w becomes infinite α times at a point a, β times at a point b, and h times at the point $z = \infty$. If we let

$$h + \Sigma\alpha + \Sigma\beta = m,$$

then SZ is an integral function of z of the mth degree. Attending now to the quantity σ, we see that SZ is also an integral function of σ of the nth degree. Therefore, if we suppose SZ to be arranged according to powers of σ, we can say that SZ is an integral function of σ of the nth degree, the coefficients of which are integral functions of z at most of the mth degree; this Riemann was in the habit of expressing by the symbol

$$F\left(\begin{smallmatrix}n & m\\ \sigma, & z\end{smallmatrix}\right).$$

This expression vanishes when σ acquires one of the values $w_1, w_2, \cdots, w_n$, and hence these are the n roots of the equation

$$F\left(\begin{smallmatrix}n & m\\ w, & z\end{smallmatrix}\right) = 0.$$

Therefore: *An n-valued function, which becomes infinite of multiplicity m, is the root of an algebraic equation between w and z, of the nth degree with regard to w, the coefficients of which are integral functions of z at most of the mth degree.*[1]

SECTION VIII.

INTEGRALS.

A. *Integrals taken along closed lines.*

42. We proceed now to complete the propositions given in Section IV., in which, however, we will consider only infinite values of finite order. According to the principles relating to infinite values of functions established in the preceding section, we can express the proposition derived in § 20 in the form: If the integral

$$\int f(z)dz$$

be taken along a closed line enclosing only one point of discontinuity a, which is not a branch-point, and at which $f(z)$ becomes infinite of the first order, then

$$\int f(z)dz = 2\pi i \lim \left[(z-a)f(z)\right]_{z=a}.$$

---

[1] We observe that here the coefficient of the highest power of w is not necessarily unity, as was assumed in § 38.

We will now investigate the value of this integral, when $f(z)$ is infinite of the nth order at a. By § 29 we have in the domain of the point a

$$(1) \quad f(z) = \frac{c'}{z-a} + \frac{c''}{(z-a)^2} + \cdots + \frac{c^{(k)}}{(z-a)^k} + \cdots + \frac{c^{(n)}}{(z-a)^n} + \psi(z),$$

wherein $\psi(z)$ remains finite and continuous for $z = a$. If we now construct $\int f(z)dz$ in reference to a closed line round the point a, we can choose for that purpose an arbitrarily small circle described round a, and we then have first

$$\int \psi(z)dz = 0,$$

and in addition $\quad \int \dfrac{c'dz}{z-a} = 2\pi i c'.$

Next, letting $\quad z - a = r(\cos\phi + i\sin\phi),$

we get $\quad \displaystyle\int \frac{c^{(k+1)}dz}{(c-a)^{k+1}} = \frac{c^{(k+1)}i}{r^k} \int_0^{2\pi}(\cos k\phi - i\sin k\phi)d\phi.$

But this integral vanishes, because for every integral value of k not zero

$$\int_0^{2\pi}\cos k\phi \, d\phi = 0, \quad \int_0^{2\pi}\sin k\phi \, d\phi = 0.$$

Therefore for every value of k different from unity

$$\int \frac{c^{(k)}dz}{(z-a)^k} = 0.$$

Therefore, in the integration, all terms except the first vanish from expression (1) and we have

$$\int f(z)dz = 2\pi i c'.$$

Accordingly the integral is always equal to zero, if the term $\dfrac{c'}{z-a}$ be wanting in the expression which defines the nature of the infinite value of $f(z)$; but if this term be present, the integral has the value $2\pi i c'$.

Let us proceed now to the case of a branch-point. If b be a point of discontinuity at which m sheets of the z-surface are connected, we have in the vicinity of the point b (§ 38)

$$(2) \quad f(z) = \frac{g'}{(z-b)^{\frac{1}{m}}} + \frac{g''}{(z-b)^{\frac{2}{m}}} + \cdots + \frac{g^{(m)}}{z-b} + \cdots + \frac{g^{(k)}}{(z-b)^{\frac{k}{m}}} + \cdots$$

$$+ \frac{g^{(n)}}{(z-b)^{\frac{n}{m}}} + \psi(z),$$

wherein $\psi(z)$ is finite and continuous for $z = b$. If we now construct $\int f(z)dz$, taken round a closed line enclosing the point b, we can for that purpose choose an arbitrarily small circle, the circumference of which, however, must be described m times in order that it may be closed. Again in the first place

$$\int \psi(z)dz = 0,$$

and further, for

$$z - b = r(\cos \phi + i \sin \phi),$$

$$\int \frac{g^{(m)}dz}{z-b} = \int_0^{2m\pi} g^{(m)} i d\phi = 2 m\pi i g^{(m)}.$$

Finally, k denoting an integer different from m,

$$\int \frac{g^{(k)}dz}{(z-b)^{\frac{k}{m}}} = g^{(k)} \int \frac{dz}{(z-b)^{\frac{k-m}{m}}(z-b)}$$

$$= ig^{(k)} r^{\frac{m-k}{m}} \int_0^{2m\pi} \left(\cos \frac{k-m}{m}\phi - i \sin \frac{k-m}{m}\phi \right) d\phi.$$

But now again

$$\int_0^{2m\pi} \cos \frac{k-m}{m} \phi d\phi = 0, \quad \int_0^{2m\pi} \sin \frac{k-m}{m} \phi d\phi = 0,$$

as long as k is not equal to m, and hence, also,

$$\int \frac{g^{(k)}dz}{(z-b)^{\frac{k}{m}}} = 0.$$

Therefore, in the integration of expression (2), all the terms, with the exception of $\dfrac{g^{(m)}}{z-b}$, vanish, and consequently

$$\int f(z)dz = 2\,m\pi i g^{(m)}.$$

Therefore, this integral, likewise, always vanishes when the term $\dfrac{g^{(m)}}{z-b}$ is wanting in the expression which defines the nature of the infinite value of $f(z)$, and the proposition in general can be expressed in the form: —

The integral $\int f(z)dz$, taken round a point of discontinuity about which the z-surface winds m times, and at which f(z) becomes infinite of a finite order, has a value different from zero, when, and only when, the term which becomes infinite of the first order is present in the expression defining the nature of the infinite value of f(z); and this value is equal to $2\,m\pi i$ times the coefficient of this term. If the point of discontinuity be not a branch-point, we have only to let $m = 1$.

43. In the consideration of the infinite value of z, we have to conceive the infinite extent of the plane, by § 14, as a sphere with an infinite radius, therefore as a closed surface, and to imagine the value $z = \infty$ to be represented by a definite point. We can then also speak of closed lines which enclose the infinitely distant point. We will now investigate the behavior of integrals when they are taken round such closed lines. These still form closed lines when we imagine the infinite sphere again extended in the plane, but then that region which contains the point $z = \infty$ lies in the plane outside the line by which it is bounded.

If another variable u be introduced instead of z, by letting

$$z - h = \frac{1}{u-k}$$

and
$$f(z) = \phi(u),$$

wherein h and k may denote two points to be chosen arbitrarily, then to every point z corresponds a point u, and conversely.

But to the points $z = h$ and $u = k$ correspond respectively $u = \infty$ and $z = \infty$. If we let

$$z - h = r(\cos\phi + i\sin\phi),$$

then
$$u - k = \frac{1}{r}(\cos\phi - i\sin\phi).$$

Now, if z describe a closed line Z enclosing the point h, then ϕ increases from 0 to a multiple of 2π; hence the corresponding line U, described by u, also encloses the point k, and indeed in an equal number of circuits, but it is to be described in the opposite direction. Further, if z go from the perimeter of Z outward, then r, or modulus of $z - h$, increases; therefore $\frac{1}{r}$, the modulus of $u - k$, decreases, and hence u goes from the perimeter of U inward. Accordingly, to all points z lying without Z correspond such points u as lie within U. If we now regard the curve Z as the boundary of the portion of the surface lying on the outside, the positive boundary-direction for this is opposite to that for the part of the surface in the interior; hence Z and U are simultaneously traversed in the positive boundary-direction of corresponding portions of the surface.

Now
$$f(z) = \phi(u), \quad dz = -\frac{du}{(u-k)^2};$$
therefore we obtain
$$\int f(z)dz = -\int \frac{\phi(u)du}{(u-k)^2},$$

wherein the first integral refers to the curve Z, the second to the corresponding curve U, taken over both in the positive boundary-direction. Now, if there be in a closed surface a curve Z enclosing the point ∞, this becomes in the plane a closed line which bounds the portion of the surface lying on the outside. The arbitrarily assumed point h can always be so chosen that it lies within the curve Z; then the part of the surface containing the point $z = \infty$ corresponds to the part of the surface lying within U, and the above equation

$$\int f(z)dz = -\int \frac{\phi(u)du}{(u-k)^2}$$

INTEGRALS. 177

holds, extended along the positive boundaries of these portions of the surface. Thus the value of the boundary integral $\int f(z)dz$ depends upon the nature of the function $\dfrac{\phi(u)}{(u-k)^2}$. We now need consider only such curves Z as contain no points of discontinuity, $z = \infty$ excepted; then $\phi(u)$ becomes infinite within U at most for $u = k$. Thus the inquiry comes to this, whether and how $\dfrac{\phi(u)}{(u-k)^2}$ is infinite for $u = k$. This expression is equal to $(z-h)^2 f(z)$, and since for $z = \infty$

$$\lim (z-h)^2 f(z) = \lim z^2 f(z),$$

it follows that, *not so much the nature of the function $f(z)$ at the point $z = \infty$, as much more that of the function $z^2 f(z)$, is serviceable for the evaluation of the boundary integral.* But if this principle be observed, all previous propositions which hold for boundary integrals are valid also for such closed lines as enclose the point ∞; at the same time it is to be kept in view, however, that when the integral is taken in the positive boundary-direction of the piece of the surface containing the point ∞, the value of the integral must have the opposite sign. Therefore, if $z^2 f(z)$ be finite for $z = \infty$, that is, if $\lim [zf(z)]_{z=\infty} = 0$, the integral is zero; hence it does not suffice for this end that $f(z)$ remain finite, the function must rather be infinitesimal of the second order. Further, if $z^2 f(z)$ be infinite of the first order, that is, if $\lim [zf(z)]_{z=\infty}$ be finite and not zero, then

$$\int f(z)dz = -2\pi i \lim [zf(z)]_{z=\infty},$$

the integral being taken in the positive boundary-direction round the point ∞. In general, the integral has a value different from zero when, and only when, in the development of $f(z)$, in ascending and descending powers of z, a term of the form $\dfrac{g}{z}$ is present.

Ex. 1.
$$\int \frac{dz}{1+z^2};$$

here $\lim [zf(z)]_{z=\infty} = \lim \left[\frac{z}{1+z^2}\right]_{z=\infty} = 0,$

therefore the integral, taken along a line enclosing the point ∞, is equal to zero. In fact, each line enclosing the two points $z = -i$ and $z = +i$ is at the same time a line enclosing the point ∞, since the function has no other points of discontinuity, and we have already seen (§ 20) that for such a line the integral has the value zero.

Ex. 2. If the integral

$$\int \frac{dz}{z}$$

be taken along a line round the origin in the direction of increasing angles, it has the value $2\pi i$. But the same line is also one which encloses the point ∞, since the function $f(z) = \frac{1}{z}$ possesses only the one point of discontinuity $z = 0$. Now, although in this case $f(z)$ is not infinite for $z = \infty$, yet the integral has a value different from zero, because

$$\lim [zf(z)]_{z=\infty} = \lim \left(z \cdot \frac{1}{z}\right)_{z=\infty} = 1.$$

We therefore obtain

$$\int \frac{dz}{z} = -2\pi i;$$

and in fact the line must be described in the opposite direction if it bound in a positive direction the part containing the point ∞.

Ex. 3. We can from this principle find the value of

$$J = \int \frac{dz}{\sqrt{1-z^2}},$$

extended along a line running in the first sheet, which encloses the two discontinuity- and branch-points $+1$ and -1, these being joined by one branch-cut. For such a line encloses at the same time the point ∞, without including any other point of discontinuity.

In this example
$$\lim [zf(z)]_{z=\infty} = \lim\left[\frac{z}{\sqrt{1-z^2}}\right]_{z=\infty} = \pm\frac{1}{i};$$
therefore
$$J = \pm 2\pi,$$

where the sign is yet to be determined. But, on the other hand, the line which encloses the points $+1$ and -1 can be contracted up to the branch-cut. If we then agree that the radical is to have the sign $+$ on the left side of the branch-cut (taken in the direction from -1 to $+1$) in the first sheet, and hence the sign $-$ on the right side of the same, likewise in the first sheet (cf. § 13), then also

$$J = \int_{-1}^{+1}\frac{dz}{\sqrt{1-z^2}} - \int_{+1}^{-1}\frac{dz}{\sqrt{1-z^2}} = 2\int_{-1}^{+1}\frac{dz}{\sqrt{1-z^2}}$$
$$= 4\int_0^1\frac{dz}{\sqrt{1-z^2}},$$

integrated in the direction of decreasing angles (in the positive boundary-direction round the point ∞). Since in this case all the elements of the integral are positive, J must also be positive, and hence

$$J = 4\int_0^1\frac{dz}{\sqrt{1-z^2}} = +2\pi,$$

and therefore also
$$\int_0^1\frac{dz}{\sqrt{1-z^2}} = \frac{\pi}{2}.$$

With respect to the circumstance that the integral preserves a finite value, although the function $\dfrac{1}{\sqrt{1-z^2}}$ becomes infinite for $z = 1$, compare the following paragraphs.

B. *Integrals along open lines. Indefinite integrals.*

44. We will inquire in this paragraph whether and under what conditions a function defined by an integral may remain finite, when the upper limit of the same either acquires a value for which the function under the integral sign becomes infinite, or the limit itself tends towards infinity. We will inquire further in what manner the function defined by the integral becomes infinite, if it do not remain finite in these cases. But at the same time we limit ourselves to such integrals as contain algebraic functions under the integral sign.

Let
$$F(t) = \int_h^t \phi(z)dz$$

be the integral to be investigated, wherein h denotes an arbitrary constant. We here consider only such paths of integration as lead to the same value of the function; the next section will show that the multiformity of a function defined by an integral, arising from different paths of integration, does not affect the considerations here employed.

If we assume in the first place that $\phi(z)$ becomes infinite of the nth order at a point $z = a$, which is not a branch-point, we can, by § 29, let

(1) $\quad \phi(z) = \dfrac{c'}{z-a} + \dfrac{c''}{(z-a)^2} + \cdots + \dfrac{c^{(n)}}{(z-a)^n} + \psi(z),$

wherein $\psi(z)$ remains finite for $z = a$. From this follows

$$\int_h^t \phi(z)dz = c' \int_h^t \frac{dz}{z-a} + c'' \int_h^t \frac{dz}{(z-a)^2} + \cdots + c^{(n)} \int_h^t \frac{dz}{(z-a)^n} + \int_h^t \psi(z)dz.$$

This last term is a function which also remains finite for $t = a$; if we denote it by $\lambda(t)$, and if we suppose included in

INTEGRALS. 181

it the constant terms arising from the lower limits h of the integrals, we then obtain

$$F(t) = c' \log(t-a) - \frac{c''}{t-a} - \frac{c'''}{2(t-a)^2} - \cdots - \frac{c^{(n)}}{(n-1)(t-a)^{n-1}} + \lambda(t).$$

Now, if we let the path of integration end in the point $t = a$, then the function defined by the integral is distinguished from a function $\lambda(t)$, which remains finite for $t = a$, by a quantity which contains the term $\log(t-a)$. We say in this case, $F(t)$ *becomes logarithmically infinite*. This case occurs when in the expression (1) for $\phi(z)$ the term $\dfrac{c'}{z-a}$ is present. If, on the other hand, this term be wanting, the logarithm drops out and $F(t)$ becomes infinite of an integral order. But, finally, $F(t)$ remains finite for $t = a$, only when

$$\lim \left[(z-a)\phi(z)\right]_{z=a} = 0;$$

that is, when $\phi(z)$ itself remains finite for $z = a$.

Next, let us assume that the point of discontinuity a is at the same time a branch-point. If m sheets of the z-surface be connected at this point, we can, by § 38, let

$$(2) \quad \phi(z) = \frac{g'}{(z-a)^{\frac{1}{m}}} + \frac{g''}{(z-a)^{\frac{2}{m}}} + \cdots + \frac{g^{(m)}}{(z-a)} + \frac{g^{(m+1)}}{(z-a)^{\frac{m+1}{m}}} + \cdots + \psi(z),$$

wherein $\psi(z)$ remains finite for $z = a$. From this we obtain

$$F(t) = g' \frac{m}{m-1}(t-a)^{\frac{m-1}{m}} + g'' \frac{m}{m-2}(t-a)^{\frac{m-2}{m}} + \cdots + g^{(m-1)}m(t-a)^{\frac{1}{m}} + g^{(m)} \log(t-a) - \frac{g^{(m+1)}m}{(t-a)^{\frac{1}{m}}} - \cdots + \lambda(t),$$

if, as above, $\lambda(t)$ denote the last term, which remains finite, including the constants arising from the lower limits h.

If at most the first $m-1$ terms be present in this expression, $F(t)$ remains finite for $t = a$. This case occurs when in (2) also at most the first $m-1$ terms are present. Then $\phi(t)$ is at most infinite of the order $\dfrac{m-1}{m}$, and, therefore,

$$\lim \left[(z-a)\phi(z)\right]_{z=a} = 0.$$

Consequently the condition that $F(t)$ may remain finite is here the same as before,[1] and the general propositions follow: —

(i.) *The function defined by the integral*

$$F(t) = \int_h^t \phi(z)dz$$

of an algebraic function $\phi(z)$ has a finite value for $t = a$, when, and only when,

$$\lim \left[(z-a)\phi(z)\right]_{z=a} = 0.$$

(ii.) *If $\lim \left[(z-a)\phi(z)\right]_{z=a}$ be finite and different from zero, then $F(t)$ is logarithmically infinite for $t = a$.*

(iii.) *If $\lim \left[(z-a)^\mu \phi(z)\right]_{z=a}$ have a finite value different from zero for an integral or fractional exponent μ, which is greater than unity, then $F(t)$ is infinite of an integral or fractional order; and if in the development of $\phi(z)$ the term of the form $\dfrac{g}{z-a}$ be present, $F(t)$ is at the same time logarithmically infinite.*

45. We have now to examine the value $t = \infty$. By the substitution already so often used

$$z = \frac{1}{u},$$

we reduce this case to the former. Let

$$\frac{1}{t} = \tau, \quad F(t) = F_1(\tau), \quad \phi(z) = \phi_1(u);$$

then
$$F(t) = \int_h^t \phi(z)dz = -\int_{\frac{1}{h}}^\tau \frac{\phi_1(u)du}{u^2} = F_1(\tau).$$

---

[1] We note in particular: If the function ϕ become infinite of multiplicity a at a branch-point of the $(m-1)$th order, and $a < m$, the integral remains finite.

The nature of the function $F_1(\tau)$ depends, therefore, upon the nature of the function $\dfrac{\phi_1(u)}{u^2}$ for the value $u = 0$. The results of the preceding paragraph then give:—

(1) $F_1(\tau)$ is finite, when
$$\lim\left[\frac{u\phi_1(u)}{u^2}\right]_{u\doteq 0} = \lim\left[\frac{\phi_1(u)}{u}\right]_{u\doteq 0} = 0.$$

(2) $F_1(\tau)$ is logarithmically infinite, when
$$\lim\left[\frac{\phi_1(u)}{u}\right]_{u\doteq 0}$$
is finite and not zero.

(3) $F_1(\tau)$ is of an integral or fractional order (or also at the same time logarithmically) infinite, when, for $\mu > 1$,
$$\lim\left[\frac{u^\mu \phi_1(u)}{u^2}\right]_{u\doteq 0}, = \lim\left[u^{\mu-2}\phi_1(u)\right]_{u\doteq 0}$$
is finite and not zero.

Therefore we conclude, for $t = \infty$:—

(i.) $F(t)$ *is finite, when* $\lim\left[z\phi(z)\right]_{z\doteq\infty} = 0$.

(ii.) $F(t)$ *is logarithmically infinite, when* $\lim\left[z\phi(z)\right]_{z\doteq\infty}$ *is finite and not zero.*

(iii.) $F(t)$ *is of an integral or fractional order (or also at the same time logarithmically) infinite when* $\lim\left[\dfrac{\phi(z)}{z^{\mu-2}}\right]_{z\doteq\infty}$ *is finite and not zero* (μ *positive and* >1).

Examples:—

$\displaystyle\int_0^t \frac{dz}{1+z^2}$ is logarithmically infinite for $t = \pm i$, but remains finite for $t = \infty$.

$\displaystyle\int_0^t \frac{dz}{\sqrt{1-z^2}}$ remains finite for $t = \pm 1$, and is logarithmically infinite for $t = \infty$.

$\int_0^t \dfrac{dz}{\sqrt{(1-z^2)(1-k^2z^2)}}$ remains finite for $t = \pm 1$ and $t = \pm \dfrac{1}{k}$, and also for $t = \infty$; hence it is finite for every value of t.

$\int_0^t \sqrt{\dfrac{1-k^2z^2}{1-z^2}}\, dz$ remains finite for $t = \pm 1$, and becomes infinite of the first order for $t = \infty$.

$\int_0^t \dfrac{dz}{(1-a^2z^2)\sqrt{(1-z^2)(1-k^2z^2)}}$ remains finite for $t = \pm 1$, and $t = \pm \dfrac{1}{k}$, likewise for $t = \infty$, and becomes logarithmically infinite for $t = \pm \dfrac{1}{a}$.

SECTION IX.

SIMPLY AND MULTIPLY CONNECTED SURFACES.

46. For the investigation of the multiformity of a function defined by an integral, $\int f(z)dz$, the character of the connection of the z-surface, for the function $f(z)$ under the integral sign, is of special importance. In this relation, we have already pointed out (§ 18) the marked distinction existing between those surfaces in which every closed line[1] forms by itself alone the complete boundary of a portion of the surface, and those in which every closed line does not possess this property.

We call, after Riemann, surfaces of the first kind *simply connected*, those of the second kind *multiply connected*. A circular surface, for instance, is simply connected; so is the surface of an ellipse, and in general every surface which consists of a single sheet and is bounded by a line returning

---

[1] For the present we shall understand by a closed line such a one as returns simply into itself without crossing itself.

SIMPLY AND MULTIPLY CONNECTED SURFACES. 185

simply into itself without crossing itself. Multiply connected surfaces can arise when points of discontinuity are excluded from simply connected surfaces by means of small circles. For instance, if we exclude from a circular surface a point of discontinuity a, by enclosing it in a small circle k, a surface is formed which is no longer simply connected; for, if a closed line m be drawn round k, this line does not constitute the complete boundary of a portion of the surface by itself alone, but only in connection with either the small circle k or the outer circle n. But, without excluding any isolated points, we may have multiply connected surfaces, when, for instance, they possess branch-points, and hence consist of several sheets continuing one into another over the branch-cuts.

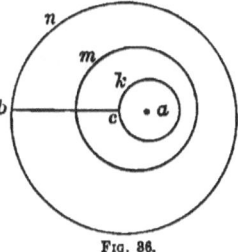

Fig. 86.

The investigations which now follow relate both to Riemann surfaces and also to other quite arbitrarily formed surfaces. Nevertheless, we must exclude such surfaces as either separate along a line into several sheets, or consist of several portions connected only in isolated points without winding round such points, as the Riemann surfaces do, by means of branch-cuts. For surfaces of this kind (divided surfaces), the properties to be developed would not be valid in their full extent. But since we have here, nevertheless, to do with surfaces the structures of which can be extraordinarily manifold, we must seek to base our investigations as much as possible upon general considerations.

In the first place, it is important to obtain a definite criterion by which we can distinguish whether or not a closed line forms by itself alone the complete boundary of a portion of the surface. To this end, we remark that two portions of a surface are said to be *connected* when, from any point of one portion to any point of the other, we can pass along a continuous line without crossing a boundary-line; in the opposite case, the portions of the surface are said to be *distinct*. If a portion A of the surface be completely bounded, it must be separated by its boundary from the other portion B of the surface; otherwise, we could pass from A to B without crossing the boundary of A, and hence that boundary would not be complete.

We will assume that the surface to be considered is bounded by one or more lines, that it possesses an *edge* consisting of one or more *boundary-edges*. We will always assume that *these lines return simply into themselves and nowhere branch.* In the Riemann surfaces, this is always the case, since in these a boundary-line has *only one* definite continuation at every place, even where it passes into another sheet; but in divided surfaces this would not always hold. In order that, within such a surface, a closed line m may form by itself alone the complete boundary of a portion of the surface, the following condition is necessary and sufficient. By the line m a piece, containing none of the original boundary-lines, must be separated from the given surface. We can now show that this condition is satisfied, when we can come from any point of the line m to the edge of the surface, without crossing the line m, only on one side; that, on the contrary, when this is possible on each side of the line m, the latter cannot form a complete boundary.

For, if we suppose the surface actually cut along the line m, two cases are possible: either the surface is divided by the section into distinct pieces, or it is not. In the latter case, no part is separated from the surface, and therefore m cannot form the boundary of a piece. Since, however, in this case all portions of the surface are still connected, we can come from either side of m to the boundary of the surface.

If, in the opposite case, the surface be divided by the section along m into distinct pieces, it reduces to only two pieces, A and B, because an interruption of the connection has nowhere occurred along one and the same side of m. Now, either both pieces A and B can contain original boundary-lines, or only one of these pieces can. If both contain boundary-lines,

SIMPLY AND MULTIPLY CONNECTED SURFACES. 187

neither of them is bounded by m alone; in this case, we can again come from m to an edge of the surface on each side of m. If, on the other hand, only the one piece B contain one or more boundary-lines, and the other piece A not, then m forms by itself alone the complete boundary of A, and we can come to the edge of the surface only on one side of m, namely, in B, not on the other side in A. Consequently a closed line m does, or does not, actually form a complete boundary by itself alone, according as we can come from m to a boundary of the surface only on one side, or on each side of m. (Cf. Fig. 36, where we can come from an arbitrary point of the line m on the one side to the part of the boundary k, on the other side to the part of the boundary n.)

This criterion cannot be immediately applied to completely closed surfaces, which, as for instance a spherical surface, do not possess a boundary. But we can assign a boundary to such a surface by taking at any place an infinitely small circle, or what is the same, a single point as boundary. (We suppose a sheet of the surface pricked through with a needle at some point.) This point, or the circumference of the infinitely small circle, then constitutes the boundary or edge of the surface. We shall always, hereafter, suppose a closed surface bounded in this way by a point, which can, moreover, be assumed in any arbitrary place in the surface. By this means the above criterion also becomes applicable to closed surfaces. We may now adduce some examples in illustration of the preceding.

(i.) A spherical surface is simply connected. For, if we draw in it any closed line m, and assume anywhere in the surface a point x as boundary, we can always come to x from m only on the one side, never at the same time on the other side; therefore every closed line m forms by itself alone a complete boundary.

(ii.) If a surface have a branch point a, at which n sheets of the surface are connected, and if a portion of the surface be bounded by a line making n circuits round the point a, and therefore being closed, the portion of the surface so bounded is simply connected. For in whatever way we may draw therein a closed line, we can always come only on the one side of the same to the edge of the surface.

(iii.) A surface consisting of two sheets closed at infinity, and possessing two branch-points a and b (Fig. 37), which are connected by a branch-cut, is a simply connected surface. We can in this case draw only three different kinds of closed lines: such as enclose no branch-point, such as enclose one, and such as enclose two. The first and last kinds are not essentially different from each other; for, according as we regard such a line m or n as the boundary of the inner or outer portion of the surface, it encloses either both branch-points or neither. But such a line as m or n always forms a complete boundary, for we can always come from it to the arbitrarily assumed boundary-point only on the one side. A finite closed line enclosing only one branch-point, for instance a, goes twice round the same, because in crossing the branch-cut it enters the second sheet, and therefore, in order to return into the first and become closed, it must again cross the branch-cut. But then it likewise forms a complete boundary.

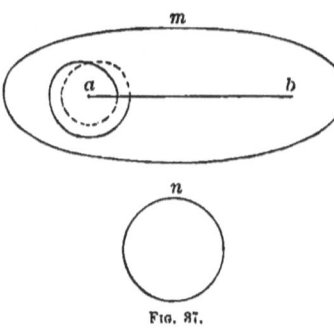

Fig. 37.

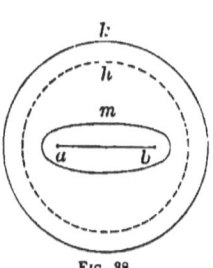

Fig. 38.

(iv.) The preceding surface becomes multiply connected, when once it is bounded in each sheet by a closed line (h and k, Fig. 38; the dotted line runs in the second sheet); for now a line enclosing a and b in the first sheet does not form a complete boundary, because we can come from it to the edge of the surface on each side; namely,

SIMPLY AND MULTIPLY CONNECTED SURFACES. 189

on the one side directly to k, on the other to h over the branch-cut.

(v.) A surface consisting of two sheets, closed at infinity, and possessing four branch-points joined in pairs by branch-cuts ab, cd, is multiply connected (Fig. 39). For, if we draw a line m, enclosing the points a and b in the first sheet, we can come from the same to the arbitrarily assumed boundary-point x on each side. If x be in the first sheet, say, this is done directly on the one side, on the other, however, by crossing the branch-cut ab. By this means we arrive in the second sheet and can, without meeting the line m (since this runs entirely in the first sheet), come to the other branch-cut cd, crossing it, we return again into the first sheet and so arrive, as before, at x.

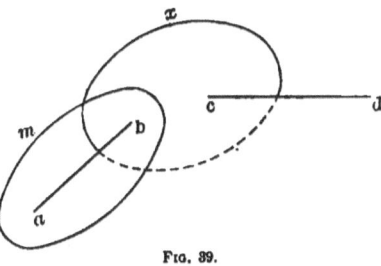

Fig. 39.

47. Now it is of the greatest importance that we be able to modify a multiply connected surface into one simply connected by adding certain boundary-lines. As will appear later (§ 56), this is always possible in a Riemann surface, if it have a finite number of sheets and branch-points, and if its boundary-lines form a finite line-system (in the meaning of § 50). These new boundary-lines are called, after Riemann, *cross-cuts*. That is, by a *cross-cut* is understood in general a line which begins at one point of a boundary, goes into the interior of the surface and, without anywhere intersecting either another boundary-line or itself, ends at a point of the boundary. In order that the meaning and extent of this definition may be made perfectly clear, let us consider somewhat more in detail the different kinds of cross-cuts. A cross-cut can connect with each other two points of the same boundary-line (ab, Fig. 40); also, two points (cd) situated on different boundary-lines. It can also

190   THEORY OF FUNCTIONS.

end in the same point of a boundary-line in which it began (*efge*), and therefore be a closed line. This is especially the case, when a cross-cut is to be drawn in a closed surface; for, since in such a surface the original boundary consists of only a single point (§ 46), the cross-cut must begin in this point and also end in it, unless the case to be immediately mentioned occurs, in which it ends in a point of its previous course. It has been stated already that the cross-cuts are to be regarded as boundary-lines, added to the already existing boundary-lines. Hence, if a cross-cut have been begun, its points are immediately regarded as belonging to a newly added boundary; and since it is only necessary for a cross-cut to end in a point of a boundary, it can also end in one of its previous points (*abcd*, Fig. 41).

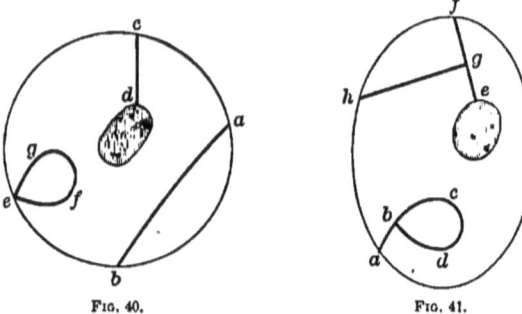

Fig. 40.   Fig. 41.

For the same reason, since each cross-cut already drawn forms part of the boundary, a subsequent cross-cut can begin or end at a point of a previous one. (Fig. 41, where *ef* is a previous cross-cut, and *gh* a subsequent one.) Finally, stress is to be laid upon the following consideration. Since a cross-cut is never to cross a boundary-line, it is also never to cross a previous cross-cut. Therefore, if a line joining two boundary-points cross a previous cross-cut, such a line forms not one, but two cross-cuts; since one ends at the point of intersection, and at the same point a new one begins. Thus, for instance, in Fig. 42, the two lines *ab* and *cd* form not two, but three

SIMPLY AND MULTIPLY CONNECTED SURFACES. 191

cross-cuts; namely, according as *ab* or *cd* was first drawn, either *ab, ce, ed,* or *cd, ae, eb*. In like manner, two cross-cuts are formed by the line *fghi*, namely, *fghg* and *gi*, or *ighg* and *gf*.

In all cases a cross-cut is to be regarded as a section actually made in the surface, so that in it two boundary-lines (the two edges produced by the section) are always united, one of which belongs as boundary-line to the portion of the surface lying on one side of the cross-cut, the other to that on the other side.

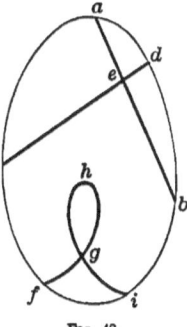

Fig. 42.

The possibility of modifying multiply connected surfaces into simply connected surfaces may be brought into consideration in the first place in some simple cases. For instance, if a cross-cut *bc* be drawn in the surface bounded by the lines *k* and *n* (Fig. 43), and if both sides of the same be included in the boundary (since the surface is regarded as actually cut through along *bc*), a closed line can no longer be drawn to include

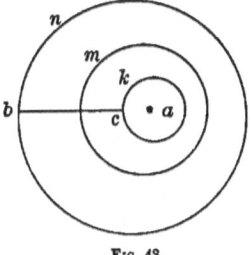

Fig. 43.

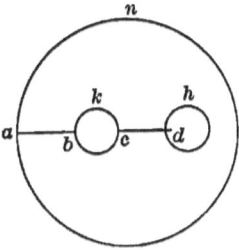

Fig. 44.

k, but every closed line forms by itself alone a complete boundary. The same condition is obtained in the surface bounded by the lines *h, k, n* (Fig. 44) by means of the cross-cuts *ab, cd*. We remark that, in the last example, the modification into a simply connected surface can be effected in

192   THEORY OF FUNCTIONS.

several ways, but always by means of two cross-cuts ab, cd; for example, as in Fig. 45 and Fig. 46.

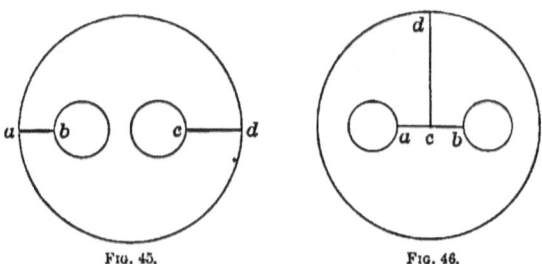

Fig. 45.   Fig. 46.

48. We now proceed to the general investigation of the resolvability of a multiply connected surface into one simply connected, and to that end we prove some preliminary propositions.

I. *If a surface T be not resolved by any cross-cut ab into distinct pieces, it is multiply connected.*

Let us first assume that the end-points a and b of the cross-cut both lie on the original boundary of T; by this assumption, however, we are not to exclude the case in which a and b coincide. Since, according to the hypothesis, the cross-cut ab does not divide the surface into distinct pieces, the two sides of the same are connected, and a closed line m can be drawn from a point c on the cross-cut which leads from one side of it through the interior of the surface to the other side.[1] Such a closed line m, however, does not form by itself alone a complete boundary; for we can come from c on each side of m along the cross-cut to the edge of T, that is, to a and b. Therefore T is in fact multiply connected. The same is true, if the cross-cut, not resolving the surface, be such a one as ends in a point of its previous course (cf. Fig. 41, $abcd$). For, in that case, we must be able to draw from a point c, situated on the closed part of the cross-cut, a closed line m leading from the one side of the

[1] This construction is to be so understood here, and likewise also later, that the line m would be closed, if the cross-cut did not exist.

same to the other side.[1] But such a closed line m does not form a complete boundary, since we can come from c on each side of m along the cross-cut to the edge of the surface, that is, to a. (In Fig. 41, the two paths are cba and $cdba$.)

II. *It is always possible to draw, in a multiply connected surface, at least one cross-cut which does not resolve the surface into distinct pieces.*

Since the surface is multiply connected, there is in it at least one closed line m which does not form by itself alone a complete boundary; thus we can come from each side of this line to the edge of the surface (§ 46). We can therefore draw from a point c of the line m two lines, ca and cb, which go on different sides of the line m through the interior of the surface and end in the points a and b of the edge (wherein a and b may also coincide). These two lines then form together a cross-cut ab, because together they may be regarded as one line which begins in a point a of the edge and, without anywhere crossing a boundary line, ends in a point b of the edge. This cross-cut does not, however, resolve the surface, for we can come along the line m itself from the one side of the cross-cut to the other side of the same; so that these two pieces of the surface are connected and not distinct.

NOTE.—The foregoing shows at the same time how we can draw in a multiply connected surface a cross-cut which does not divide the surface, when we know in the surface a closed line which forms by itself alone a complete boundary.

III. *A surface consisting of one piece can be resolved at most into two pieces by a cross-cut.*

Either the portions of the surface lying on each side of the cross-cut are connected, in which case the cross-cut does not resolve the surface; or they are not connected, in which case those portions of the surface lie in distinct pieces. If the number of the latter amount to more than two, there must

---

[1] This is possible when, for instance, within the closed portion bcd, there is a branch-cut which the line m can cross, thereby coming into another sheet in which it does not meet the cross-cut, and when by means of a second branch-cut it can return to the initial point.

occur an interruption of the connection in a portion of the surface lying on one and the same side of the cross-cut; but this is not the case, because the cross-cut nowhere crosses a boundary-line.[1]

IV. *A simply connected surface is resolved by every cross-cut into two distinct pieces, each of which is again by itself simply connected.*

The first part of the proposition follows immediately from I. and III.; for, if no cross-cut would resolve the surface, the latter could not be simply connected, and it cannot be resolved into more than two pieces. But it is evident without further proof that each of the pieces formed must be by itself simply connected; for, since in the unresolved surface every closed line forms by itself a complete boundary, the same also holds of every closed line which runs entirely within one of the pieces formed.

V. *A simply connected surface is resolved by q cross-cuts into $q + 1$ distinct pieces, each of which is by itself simply connected.*

If, after the surface is first resolved into two distinct pieces by one cross-cut (IV.), a new cross-cut be drawn, this can run in only one of the two pieces already formed, because it is not allowed to cross the first cross-cut (§ 47); but it resolves the portion in which it falls into two pieces, so that the two cross-cuts divide the surface into three pieces. These again are each by itself simply connected. If we now draw a new cross-cut, again only one of the already existing pieces is resolved; and likewise the number of pieces is increased only by unity for every succeeding cross-cut. Therefore, at the end, after q cross-cuts have been drawn, $q + 1$ distinct pieces are formed; and these are each by itself simply connected (IV.).

Cor. From this follows immediately: If there be a system of surfaces consisting of a distinct pieces, each by itself simply connected, this is resolved by q cross-cuts into $a + q$ simply connected pieces.

---

[1] In the case of a cross-cut ending in a point of its previous course, the one side consists of the portions of the surface on the inside contiguous to the closed part, the other side of the remaining portions of the surface.

NOTE. — The foregoing considerations are also applicable to the case in which the a pieces of which the original system consists are not all simply connected. In that case, however, the distinction occurs, that there may now be cross-cuts not resolving the pieces in which they are constructed; and finally, as a consequence, that, after the introduction of the q cross-cuts, less than $a+q$ distinct pieces are formed. But more than $a+q$ pieces cannot arise. We conclude therefore: If q cross-cuts be drawn in a system of surfaces consisting of a pieces, the number of pieces arising thereby is either equal to or less than $a + q$, but never greater than $a + q$.

VI. *If a surface be resolved by every cross-cut into distinct pieces, it is simply connected.*

For, were it multiply connected, it could not be resolved by every cross-cut (II.).

VII. *If a surface T be resolved by any one definite cross-cut Q into two distinct pieces, A and B, each of which is by itself simply connected, T is also simply connected.*

We shall show that, under the given hypothesis, every cross-cut drawn in T must resolve this surface. In the first place, it is evident that every cross-cut which lies entirely within A or B, and which therefore does not cross Q, resolves the surface; for, if such a cross-cut lie entirely within A, for instance, it resolves A into two distinct pieces (IV.), and that one of these pieces which is contiguous to Q, together with B, forms one piece of T, and the other forms a second piece distinct from the former. If, however, a cross-cut Q' cross Q one or more times, it is divided by the points of intersection into parts which form cross-cuts in either A or B, and which therefore again resolve these portions into distinct pieces (IV.). Thus we cannot come from the one side of Q' to the other side either in A or in B. But then this is also not possible in T, *i.e.*, by crossing Q, because thereby we always come only from A to B, or conversely. Therefore Q' likewise divides the surface. Since this is resolved into two distinct pieces by every cross-cut, it is simply connected (VI.).

VIII. *If a multiply connected surface be resolved into two distinct pieces by a cross-cut, at least one of the pieces is again multiply connected.*

For, if both were simply connected, T could not be multiply connected (VII.).

IX. *If a surface consisting of one piece be resolved by q cross-cuts into $q + 1$ pieces, each of which is by itself simply connected, then it is itself simply connected.*

Each of the cross-cuts drawn divides the part in which it falls into two distinct pieces; for, if only a single one should not do this, there would at the end be less than $q + 1$ distinct pieces, since a cross-cut can never divide a portion into more than two pieces (III.). If the given surface were multiply connected, the first cross-cut could at most cut off one simply connected piece (VIII.), the other piece remaining multiply connected. If we now assume, in order to emphasize the most favorable case, that the cross-cut is drawn every time in the multiply connected portion, so that from this a simply connected piece is detached, at the end a piece would remain which is not simply connected. In general, by each mode of resolution, either less than $q + 1$ pieces would have been formed, or at least one of these pieces would necessarily be multiply connected.

49. From these preliminary principles we now proceed to the following fundamental proposition: —

If a surface, or a system of surfaces, T, be resolved in one way by q_1 cross-cuts Q_1 into a_1 distinct pieces, and in a second way by q_2 cross-cuts Q_2 into a_2 distinct pieces, in such a manner that both the a_1 pieces of the first way and also the a_2 pieces of the second way are, each by itself, simply connected, then in all cases

$$q_1 - a_1 = q_2 - a_2.$$

Proof.[1] — The two systems of surfaces formed from T by means of the cross-cuts Q_1 and Q_2 may be called T_1 and T_2 respectively. If we draw either the lines Q_2 in T_1, or the lines Q_1 in T_2, we obtain in both cases the same system of surfaces, exactly the same figure. Call this new system of surfaces $\mathfrak{T}$.

[1] Riemann, *Grundlagen*, u. s. w., s. 6.

The lines Q_2 form, it is true, q_2 cross-cuts in the original surface T, but not necessarily the same number when drawn in T_1; for, since on the one hand the lines Q_2 cease to exist as cross-cuts in T_1 if they coincide entirely with the lines Q_1, and since on the other hand they may be divided by the lines Q_1 into several parts (each part forming a distinct cross-cut), the number of cross-cuts actually formed in T_1 by the lines Q_2 may be less, or even greater than q_2. Likewise, also, the number of cross-cuts formed by the lines Q_1 in the system T_2 may be different from q_1. We will designate the cross-cuts formed by the lines Q_2 in T_1 by Q_2', their number by q_2'; the cross-cuts formed by the lines Q_1 in T_2 by Q_1', their number by q_1'. The essential feature of the proof consists then in this, that, if we let

$$q_2' = q_2 + m,$$

then also $\qquad q_1' = q_1 + m.$

To prove this, let us direct our attention to the end-points[1] of the cross-cuts, observing that the number of cross-cuts is half as great as the number of their end-points, and that this is invariably the case if we only count twice a point at which the initial point of one cross-cut coincides with the terminal point of another. Accordingly, the number of end-points of the q_2 cross-cuts Q_2 is $2 q_2$. But if these be regarded as cross-cuts Q_2' in the system T_1, already resolved by the lines Q_1, on the one hand some end-points of the lines Q_2 may cease to be end-points of the lines Q_2', and on the other hand new points may occur as end-points of the lines Q_2'. (Cf. Fig. 47, wherein the lines Q_1 are represented by the heavier lines, the lines Q_2 by the lighter. In places where a line Q_1 coincides with a line Q_2, wholly or in part, they are represented running closely beside each other.)

(1) An end-point of a line Q_2 is always at the same time an end point of a line Q_2', if it do not fall on one of the lines Q_1 (*e.g.*, α or β), and also in the case when only an end-*point* of a

[1] We will call the initial and terminal points of a cross-cut together the two end-points of the same.

198   THEORY OF FUNCTIONS.

line Q_2 coincides with an end-*point* of a line Q_1 (*e.g.*, *c* or *g*). On the other hand an end-*point* of a line Q_2 is *not* at the same time also an end-point of a line Q_2', if the line Q_2 coincide for a distance from this end-point (or also completely) with a line Q_1 (*e.g.*, *ds*, *δr*, *op*, *po*). In such a case the point in question either ceases altogether to exist as an end-point of a line Q_2' (*e.g.*, *o* or *p*), or it is only displaced as such. (While, for instance, the cross-cut $ds\beta$, regarded as a line Q_2, begins at *d*, the same, regarded as a line Q_2', begins only at *s*; *d* is therefore an end-point of a line Q_2, but not of a line Q_2'.) Now this can occur in two cases: either the segments which coincide are both end-pieces of a line Q_2 and a line Q_1 respectively (*e.g.*, *ds*), or an end-piece of a line Q_2 coincides with a

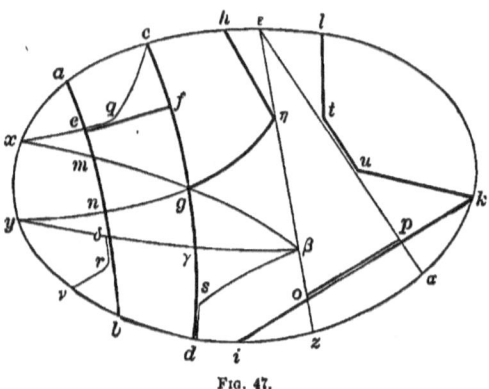

Fig. 47.

mid-piece of a line Q_1 (*e.g.*, *δr*, *op*, *po*). If, therefore, ν be the number of times that an end-piece of a line Q_2 coincides with an end-piece of a line Q_1, and ν_2 be the number of times that an end-piece of a line Q_2 coincides with a mid-piece of a line Q_1, then

$$\nu + \nu_2$$

is the number of end-points of the lines Q_2 which are not at the same time end-points of the lines Q_2'. The number $2 q_2$ of end-points of the lines Q_2 must therefore be diminished by $\nu + \nu_2$.

SIMPLY AND MULTIPLY CONNECTED SURFACES. 199

Similar considerations are applicable to the cross-cuts Q_1. An end-point of a line Q_1 ceases to be an end-point of a line Q_1', when an end-piece of a line Q_1 coincides either with an end-piece of a line Q_2 (*e.g.*, ds) or with a mid-piece of a line Q_2 (*e.g.*, eq). If therefore ν_1 be the number of times that an end-piece of a line Q_1 coincides with a mid-piece of a line Q_2, then

$$\nu + \nu_1$$

is the number of end-points of the lines Q_1 which are not at the same time end-points of the lines Q_1'; therefore the number $2q_1$ of end-points of the lines Q_1 must be diminished by $\nu + \nu_1$.

(2) But now new points appear as end-points of the lines Q_2' or Q_1', which are not end-points of the lines Q_2 or Q_1. Let us again consider first the lines Q_2'. As new end-points of these lines appear, in the first place, the displaced points mentioned above (*e.g.*, r or s); then, also, both those points at which a *mid-point* of a line Q_2 coincides with a *mid-point* of a line Q_1 (*e.g.*, γ and η), and those near which a *mid-piece* of a line Q_2 coincides with a *mid-piece* of a line Q_1 (*e.g.*, tu). All these cases can be characterized as those in which the lines Q_1 and Q_2 either meet or separate in their mid-course. Let μ denote the number of times that this occurs. The following considerations are to be noted, however, concerning the determination of this number μ. In the first place, wherever a line Q_2 has common with a line Q_1 only a single mid-point (not a segment; *e.g.*, at γ or η), this point must be counted twice, because it is a terminal point of a line Q_2' and at the same time an initial point of a new line Q_2'. We will, however, stipulate that the number μ be so determined that its value shall be independent of whether we put the lines Q_2 in relation to the lines Q_1, or, conversely, the lines Q_1 in relation to the lines Q_2. This requires us to take the greater number, whenever the one relation produces a greater number of points to be counted than the other, for two particular cross-cuts. The points which are thus counted too often must then be set aside. Now this case occurs with the cross-cuts Q_2' when, and only when, an end-

piece of a line Q_1, coinciding with a line Q_2, terminates[1] in that mid-point of the line Q_2 with which a mid-point of another line Q_1 coincides (*e.g.*, at e, where fe ends; here in the determination of μ, e must be counted twice, because xc and ab have only one point common at e, and e is at the same time an end-point of ea and em; but e occurs on xc only once as an end-point, namely, of xe, while the segment ec, regarded as a line Q_2', begins only at q, which point must likewise be counted in the determination of μ). This case occurs therefore when, and always when, an end-piece of a line Q_1 coincides with a mid-piece of a line Q_2; for, that this may be possible, another line $Q_1{}^2$ must at the same time pass through the end-point of this line Q_1. The number of times that an end-piece of a line Q_1 coincides with a mid-piece of a line Q_2 was denoted above by ν_1. Therefore, among the points counted in the determination of the number μ, are ν_1 such points as are not end-points of the lines Q_2'. Therefore the number of new end-points of the lines Q_2' which occur amounts to

$$\mu - \nu_1{}^3.$$

[1] This case can also occur when the line Q_1 ends in a point of its previous course. Cf. the third note.

[2] Cf. preceding note.

[3] There are also cases in which the number μ can be computed in different ways; the difference $\mu - \nu_1$, however, remains the same. In Fig. 48, two such cases are given. The cross-cut $abcdb$ (a line Q_1), ending in a point of its previous course, coincides with ef (as a line Q_2) along the part bd. Now, if we take the first in the sense $abdcb$, we have two points μ on each cross-cut, namely, b and d; therefore $\mu = 2$, and, bd being at the same time a mid-piece, $\nu_1 = 0$. If, on the other hand, we take the line Q_1 in the sense $abcdb$, b is to be counted twice, and we have therefore now $\mu = 3$; at the same time, however, db is an end-piece of the line Q_1 which coincides with a mid-piece of the line Q_2, and therefore $\nu_1 = 1$. The difference $\mu - \nu_1$ is in both cases the same. The other example is similar to the preceding. In this we have the choice of assuming $lhik$ as the previous cross-cut

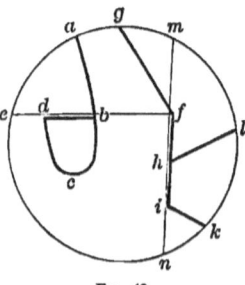

FIG. 48.

SIMPLY AND MULTIPLY CONNECTED SURFACES. 201

The reasoning is similar in the determination of the number of points which enter as new end-points of the lines Q_1'. The number μ remains the same as before, if it be determined in the manner given above. Among these μ points, however, those are not end-points of the lines Q'_1 near which an end-piece of a line Q_2 coincides with a mid-piece of a line Q_1 (e.g., near δ, which point is to be counted twice; it enters, it is true, as an end-point of δn, but not of δb, because this cross-cut, regarded as a line Q_1', begins only at r). According to the above notation this occurs ν_2 times; therefore the number of points which enter as new end-points of the cross-cuts Q_1' equals

$$\mu - \nu_2.$$

Now, according to (1), $\nu + \nu_2$ end-points are to be subtracted from the original $2q_2$ end-points, and, according to (2), $\mu - \nu_1$ new end-points are to be added. Therefore the number $2q_2'$ of end-points of the cross-cuts Q_2' is

$$2q_2' = 2q_2 - (\nu + \nu_2) + (\mu - \nu_1),$$

and hence $\quad q_2' = q_2 + \dfrac{\mu - (\nu + \nu_1 + \nu_2)}{2}.$

On the other hand, according to (1), $\nu + \nu_1$ end-points are to be subtracted from the original $2q_1$ end-points of Q_1, and according to (2), $\mu - \nu_2$ are to be added. The number $2q_1'$ of end-points of the cross-cuts Q_1' therefore is

$$2q_1' = 2q_1 - (\nu + \nu_1) + (\mu - \nu_2),$$

and hence $\quad q_1' = q_1 + \dfrac{\mu - (\nu + \nu_1 + \nu_2)}{2}.$

and hfg as the subsequent one, or $gfhik$ as the previous and hl as the subsequent cross-cut. The piece fhi coincides with the cross-cut mn of the second kind. If we choose the first order, we have to count three points, h, i, and f; therefore $\mu = 3$. At the same time, however, hf is an end-piece, and therefore $\nu_1 = 1$. On the other hand, with the second order, only f and i are to be counted, and therefore $\mu = 2$; but at the same time $\nu_1 = 0$, because fh is now a mid-piece.

Therefore, if we let

$$\frac{\mu - (\nu + \nu_1 + \nu_2)}{2} = m,$$

we have simultaneously

$$q_2' = q_2 + m \quad \text{and} \quad q_1' = q_1 + m;$$

this result was first to be proved.

Let us, before proceeding further, consider Fig. 47 in detail in reference to the above-described relations.

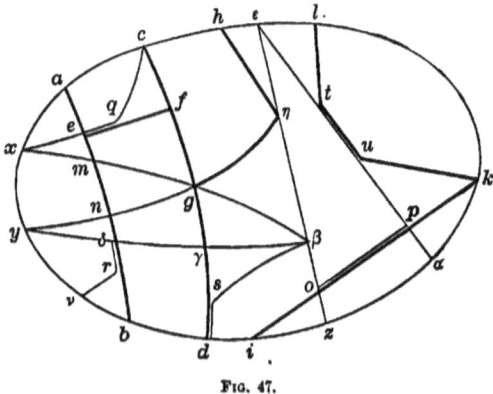

Fig. 47.

In the following table the cross-cuts Q_1 are enumerated in the first column, and, in the same row with each, are given in the second column the pieces into which it is divided if the surface be regarded as previously resolved by the lines Q_2; the latter therefore are the cross-cuts Q_1'. The columns headed Q_2 and Q_2' have similar meanings.

SIMPLY AND MULTIPLY CONNECTED SURFACES.

$Q_1.$	$Q_1'.$	$Q_2.$	$Q_2'.$
ab	ae, em, mn, nδ, rb	xc	xe, qc
cd	cg, gγ, γs	εz	εη, ηo, oz
ef	qf	xβ	xm, mg, gβ
gh	gη, ηh	yg	yn, ng
ik	io, pk	yβ	yδ, δγ, γβ
kl	ku, tl	δν	rν
		βd	βs
		εa	εt, up, pa
		op	

From this we obtain

$$q_1 = 6, \quad q_1' = 15, \quad m = 9.$$
$$q_2 = 9, \quad q_2' = 18,$$

Let us further enumerate the points μ, indicating each point which is to be counted twice by a 2 set over it:

$$\overset{2}{e}\ \overset{2}{m}\ \overset{2}{n}\ \overset{2}{\delta}\ \ \overset{2}{r}\ \overset{2}{g}\ \ \gamma\ \ s\ \ q\ \ \overset{2}{\eta}\ \overset{2}{o}\ \overset{2}{p}\ \ t\ \ u;$$

hence $\mu = 23.$

The end-pieces which coincide are:

class of $\nu \cdots ds,\qquad \nu = 1,$

" " $\nu_1 \cdots eq,\qquad \nu_1 = 1,$

" " $\nu_2 \cdots \delta r,\ op,\ po,\qquad \nu_2 = 3;$

therefore
$$m = \frac{\mu - (\nu + \nu_1 + \nu_2)}{2} = \frac{23 - 5}{2} = 9,$$
as above.

The rest of the proof is now very easily given. According to the hypothesis, the system T_1 consists of a_1 distinct pieces, each by itself simply connected. From this the system $\mathfrak{T}$ is produced by the $q_2' = q_2 + m$ cross-cuts Q_2'. The latter system consists of only simply connected pieces (§ 48, IV.). Letting 𝖀 denote the number of the latter, it follows from § 48, V. that

$$\mathfrak{U} = a_1 + q_2 + m.$$

The same system $\mathfrak{T}$ is also produced from T_2 by the

$$q_1' = q_1 + m$$

cross-cuts Q_1'. Since, however, according to the hypothesis, T_2 consists of a_2 distinct pieces, each by itself simply connected, it follows also that

$$\mathfrak{U} = a_2 + q_1 + m.$$

Therefore, $\quad a_1 + q_2 + m = a_2 + q_1 + m,$

or $\quad\quad\quad\quad q_2 - a_2 = q_1 - a_1.$

NOTE. — Having prescribed the conditions to be kept in mind in the proof of this proposition, we can prove it more briefly in the following manner:[1] Let us assume, in the first place, that the lines Q_1 and Q_2 have in common only this, that a line of the one kind simply crosses one of the other, and in such a way that the point of intersection is not an end-point of a cross-cut. If such a case occur k times, we have, according to the above explanation and with the preceding notation,

$$2\,q_2' = 2\,q_2 + 2\,k,\quad 2\,q_1' = 2\,q_1 + 2\,k,$$

and therefore $\quad q_2' = q_2 + k,\quad q_1' = q_1 + k.$

If, however, both systems lie in any arbitrary relation to each other, the lines Q_1 and Q_2 in part crossing one another at arbitrary points, in part touching one another, or even coinciding with one another wholly or in part, then an infinitely small deformation of the lines of one system can cause the coincidence either to be entirely removed or to conform only to the foregoing hypothesis. If, after such a deformation, k points of intersection occur, it follows again that

$$q_2' = q_2 + k \text{ and } q_1' = q_1 + k,$$

from which the proposition follows as above. If this hold after the infinitely small deformation, it must also be valid before the same; for by this deformation neither the number of cross-cuts nor the number of pieces into which the surface is resolved is changed.

[1] Neumann, *Vorlesungen über Riemann's Theorie*, u. s. w., S. 206.

50. The proof of the fundamental proposition given by Riemann and set forth in the preceding paragraph renders detailed discussions necessary, if none of the cases which might possibly occur are to be overlooked.

Neumann's proof is shorter, it is true, but it does not make use of the property which forms the principal part of the preceding proof, that in general

$$q_1' = q_1 + m, \quad q_2' = q_2 + m.$$

This property is important in itself, however, and will later have to be applied. On that account it is, perhaps, not unprofitable to add still another proof of the fundamental proposition, which is simpler and which yet employs the above property. Such a proof has been given by Lippich.[1] It presupposes, it is true, some knowledge of the properties of general line-systems, but it is characterized by great simplicity and leads, moreover, to a new and very important principle. To give this proof, we must introduce the following preliminary considerations.[2]

Digression on line-systems. We will consider a system of straight or curved lines which in other respects can be quite arbitrary; yet we do assume that no line of the system extends to infinity, and also that none makes an infinite number of windings in a finite region. We further assume that all the lines are connected with one another, or, if this be not the case, that they form only a finite number of pieces.

From each point of a line we can, following one or more lines, take either a single path or several paths. Accordingly we distinguish three kinds of points on the lines: (1) Let a point from which we can take only one path be called an *endpoint*; e.g., a, a_1, a_2, a_3, in Fig. 49. (2) Call a point from which we can proceed by two paths an *ordinary point;* e.g., b, b_1, b_2. (3) Let a point from which we can take more than two paths

---

[1] F. Lippich, "Bemerkung zu einem Satze aus Riemann's Theorie der Funktionen," u. s. w. (*Sitz.-Ber. d. Wien. Acad.*, Bd. 69, Abth. II., Januar 1874.)

[2] The continuation of the main investigation follows in § 52.

be called a *nodal-point*, and let it be called n-ple, when n paths proceed from it, i.e., when n line-segments meet in it. In Fig. 49, c is a triple, d a quadruple, e a quintuple nodal-point. Now if a line-system, constituted in this way, contain either no end- and nodal-points, or only a finite number of such points, it will be called a *finite line-system*. We will introduce the following additional definitions. Let a continuous line, which contains two end-points and in addition only ordinary points, be called a *simple line-segment*, e.g., aba_1; let a continuous line which contains only ordinary points be called a *simply closed line*, e.g., A.

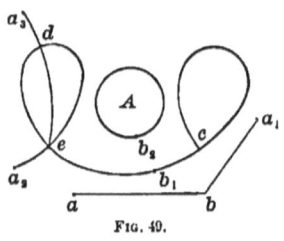

Fig. 49.

If, as will immediately occur, a description of connected lines of the system be under discussion, it will always be assumed that therein no line-segment is described more than once, but that the description is always continued and does not cease as long as a line-segment, not yet described and connected with the segment last traced, presents a path for further motion. Then, on account of the hypothesis made in regard to the construction of the line-system, the description must always come to an end sometime. This always occurs on arriving at an end-point, but at an ordinary point when, and only when, this was at the same time the initial point. At all the nodal-points, *sections* are always to be associated with the description in such a way that when a nodal-point is passed for the first time, when, therefore, we arrive at the nodal-point on one line-segment and leave it on any other, all the other line-segments meeting there are to be regarded as cut. Each of these latter, then, acquires an end-point at the nodal-point, and its connection with the other lines is regarded as broken at that point. If this method be adhered to, the description does not stop at a nodal-point when we arrive at it for the first time, but always ends when we return to it for the second time.

SIMPLY AND MULTIPLY CONNECTED SURFACES. 207

If a line-system do not contain any nodal-points, a single end-point cannot occur in it; the end-points, provided there are any, must rather occur in pairs. For if the stated description begin at one of the existing end-points, it can end neither in a nodal-point (since there are none), nor in an ordinary point (since the initial point was not an ordinary point); thus it can end only in an end-point, and, since an end of the description must sometime occur, there must still be a second end-point. Therefore, starting from an end-point, we always arrive at another end-point, and, since in the meantime we have passed through only ordinary points, we have traced a simple line-segment. Each existing end-point, therefore, in a finite line-system containing no nodal-points, is associated with a second end-point, the two together bounding a simple line-segment. Since, now, conversely, every simple line-segment also possesses two end-points, we conclude: *If a finite line-system contain no nodal-points, the number of simple line-segments constituting it is half as great as the number of existing end-points.*

In the case of an arbitrary finite line-system affected with nodal-points, we can remove the nodal-points altogether by means of the sections mentioned above, by regarding all the line-segments meeting at each nodal-point as cut except two, which are to be left connected. Then each nodal-point is changed into an ordinary point and a number of end-points; and in fact, if the point be an h-ple nodal-point, it is changed into one ordinary point and $h - 2$ end-points. Now, since the two line-segments left in connection at each nodal-point can be chosen quite arbitrarily, the sections may be effected in very many different ways. These having been effected, the line-system no longer contains nodal-points, and therefore the simple line-segments then contained in it can be found according to the preceding principle, by counting the number of end-points which occur. For instance, if we denote the number of end-points contained in the original system by e, the number of triple, quadruple, $\cdots$, n-ple nodal-points by $k_3, k_4, \cdots, k_n$, we

have, since each h-ple nodal-point furnishes $h-2$ end-points, in all
$$e + k_3 + 2k_4 + 3k_5 + \cdots + (n-2)k_n$$
simple end-points; therefore the number of simple line-segments occurring after the sections are effected equals
$$\tfrac{1}{2}[e + k_3 + 2k_4 + 3k_5 + \cdots + (n-2)k_n].[1]$$

This number depends, however, upon the originally existing end- and nodal-points, and is entirely independent of the way in which the sections were effected, which, as stated, can be made in very many different ways. We therefore conclude: *If all existing nodal-points be removed from a given finite line-system, after the sections are effected the system contains a series of simple line-segments, the number of which is constant and independent of the way in which the sections were effected.*

The line-system may originally contain simply closed lines, but the sections can always be effected in such a way that thereby no new simply closed lines are furnished. For this could only occur when, after all line-segments except three have been cut at a nodal-point, a particular one of these three is then cut. But if we choose to cut one of the two other line-segments instead of that particular one, no simply closed line will thereby be furnished. If the sections be effected in this way, and if the system contain originally no simply closed

---

[1] The number
$$e + k_3 + 2k_4 + 3k_5 + \cdots + (n-2)k_n$$
is, according to the above, an even number. If we take from it first the even number
$$2k_4 + 4k_6 + \cdots,$$
and then the even number
$$2k_5 + 4k_7 + \cdots,$$
the remaining number
$$e + k_3 + k_5 + k_7 + \cdots$$
must also be even. Thus it follows that: *In a finite line-system the number of end-points and of odd nodal-points together is always an even number.* For instance, it is not possible to draw a line-system containing, say, two end-points and a quintuple nodal-point (and no more nodal-points).

lines, after the sections are effected it consists of only simple line-segments constant in number.

It is of advantage, in a system containing no simply closed lines, to conceive the resolution into simple line-segments as effected by successive descriptions, with which, as already stated, sections are to be associated, at every nodal-point passed. Then simply closed lines can never occur, not even in the case of a triple nodal-point, because the description stops at a nodal-point only when we arrive at it for the second time. To effect the resolution, we begin the description at an end-point; then we must once more arrive at an end-point, which either originally existed or was furnished by a section at a nodal-point previously passed. We have then traced a simple line-segment. We next begin the description again, either at an originally existing end-point or at one furnished by a section, and obtain a second line-segment, etc. Should the case occur that there is, on one of the simple line-segments to be traced, no end-point at which to begin the description, then there must be a nodal-point in some part of the system not yet described; for, otherwise, this portion would contain only ordinary points, and therefore consist of only simply closed lines, while, according to the hypothesis, the system does not contain any such lines. In such a case we get an end-point at a nodal-point by cutting only one line-segment, and begin the description at this point. In doing so, if the point be a triple nodal-point, we must certainly be careful not to cut exactly that line-segment by the section of which a simply closed line could be furnished. We therefore obtain the following proposition: *A finite line-system, not containing any simply closed lines, is resolved by successive descriptions (with which a section is to be associated at every nodal-point passed) into only simple line-segments, the number of which is always the same in whatever way the description, and with it the resolution into simple line-segments, may be effected.*

51. We will now assume that in a given surface there is a finite line-system L, which satisfies the two following conditions:

(1) No end-point is to lie in the interior of the surface, but all end-points which occur are to be situated on the boundary of the surface. (If the surface be closed, it is to be regarded as having a boundary-point, by § 46.)

(2) All parts of the system L are to be connected with the boundary of the surface, so that, by following the lines, we can arrive at the boundary of the surface from any point situated on L.

Such a line-system can, it is true, contain simply closed lines, yet every simply closed line, in order to satisfy condition (2), must have at least one point in common with a boundary-line.

We now first add the boundary-lines to the system L; then every point of L situated on the boundary is a nodal-point of the system modified by the addition of the boundary-lines, since there meet in it at least three line-segments, namely, two belonging to the boundary-line, and at least one belonging to the system L. The same is also true of the boundary-point assumed in a closed surface, if we conceive this as an infinitely small boundary-line. Let us now effect the sections at all these nodal-points on the boundary in such a way that the line segments belonging to the boundary-line are left connected, but all the line-segments belonging to the system L are cut. (Thus all the line-segments meeting at a mere boundary-point must be cut.) If we next exclude again the boundary-lines, we have changed the system L into another system, which we may designate by M. Each of the points situated on the boundary in the latter system is now an end-point; and there must be at least one such point, if the system L is to satisfy condition (2). Further, the system M does not contain any closed lines, since those which may have existed in L have been removed by means of the sections. In other respects, however, the lines of M are the same as the lines of L.

Therefore, according to the last proposition of the preceding paragraph, the system M can be resolved by successive descriptions into none but simple line-segments, constant in number.

But it is now easy to show that these line-segments constitute a system of *cross-cuts* drawn in the surface. For, since from the preceding considerations there is at least one end-point situated on the boundary, the first line-segment can be traced from such a point. The second end-point, at which we then arrive, cannot by (1) be in the interior, but it is either on the boundary or it has been furnished by a section at a nodal-point passed. But in both cases the simple line-segment traced is a cross-cut; in the second case it is such a cross-cut as ends in a point of its previous course. Now there must be still another end-point (in case there are any lines not yet traced), which is situated either on the boundary or on the first cross-cut, since, otherwise, condition (2) would not be satisfied. Thus we can trace a second simple line-segment, starting from this point; the second end-point of this segment lies either on the boundary or on the first cross-cut, or it is furnished by a section on passing through a nodal-point. This second line-segment is therefore also a cross-cut. The same process is repeated as long as there remain lines not yet traced; therefore the line-segments collectively do, in fact, constitute a system of cross-cuts, and the following proposition, established by Lippich, follows: *Every finite line-system which occurs in a surface, and which satisfies conditions* (1) *and* (2), *forms a system of cross-cuts, completely determined in number, and this number remains always the same in whatever way these cross-cuts may be successively drawn.*

From the definition of cross-cuts (§ 47) it is immediately obvious that, conversely, every system of cross-cuts forms a line-system which satisfies conditions (1) and (2).

The proof of Riemann's fundamental proposition, based upon this principle, is very simple. In it we retain the notation of § 49. If we suppose first the lines Q_1 and then the lines Q_2 drawn in the surface T, we obtain a line-system which consists only of cross-cuts, and which therefore satisfies conditions (1) and (2). According to the preceding proposition this now forms at the same time a system of cross-cuts, fully determined in number, and this number may be denoted by s.

But the number of the cross-cuts Q_1 was q_1; if these be drawn, the lines Q_2 form q_2' new cross-cuts, using the former notation, and the total number of cross-cuts is thus $q_1 + q_2'$; therefore

$$q_1 + q_2' = s.$$

If now, conversely, the cross-cuts Q_2 be first drawn, the number of which was q_2, and then the lines Q_1 be added, forming q_1' new cross-cuts, then in all $q_2 + q_1'$ cross-cuts are obtained. But the line-system which now exists is exactly the same as before, except that the cross-cuts of which it consists have been drawn in a different way. Since according to the above proposition the number of these cross-cuts must nevertheless be the same, it follows that

$$q_2 + q_1' = s;$$

therefore, $q_1 + q_2' = q_2 + q_1'$ or $q_2' - q_2 = q_1' - q_1$.

If we denote the common value of these differences by m, we have

$$q'_1 = q_1 + m, \quad q_2' = q_2 + m.$$

Having established the first and principal part of the proof, we proceed exactly as in § 49.

52. Let us now consider the case in which the original surface T consists of a single connected piece, and in which, further, each of the surfaces T_1 and T_2, obtained by means of the cross-cuts Q_1 and Q_2, forms a single simply connected surface. In order that this case may occur, it is necessary in the first place that none of the cross-cuts divide the surface; therefore also, by § 48, II. and IV., that T be multiply connected and remain multiply connected, for both modes of resolution, until the next to the last cross-cut has been drawn, and be rendered simply connected only by means of the last cross-cut. In such a case $a_1 = a_2 = 1$; accordingly $q_2 - 1 = q_1 - 1$, and hence $q_2 = q_1$. From this result we obtain the following proposition:

If it be possible by means of cross-cuts to modify a multiply connected surface into one simply connected, and if this be possible in more than one way, then the number of cross-cuts by means of which the modification is effected is always the same.

In addition to this proposition the following is of the greatest importance:

If a multiply connected surface can be modified into one simply connected in any one definite way by means of q cross-cuts, then this modification is always effected by means of q cross-cuts, in whatever way these may be drawn, provided only that they do not divide the surface.

Although it has been proved in the preceding proposition that the number of cross-cuts remains the same *if* the resolution into a simply connected surface be possible in a second way, yet it remains to be discussed *whether* this resolution can in fact be effected in a second way; whether, on the contrary, for the modification into a simply connected surface, the cross-cuts must not be drawn in a definite way, according to a definite rule, so that, if they be not so drawn, the surface always remains multiply connected and a simply connected surface is never obtained, however far the drawing of the cross-cuts may be continued. But we can in fact show that this case cannot occur. We therefore assume that the surface T is modified by q cross-cuts Q_1, drawn in a definite way, into a simply connected surface T_1. Then in the first place it follows from the preceding proposition that, if instead of the former cross-cuts others be drawn, the surface T cannot be made simply connected by means of less than q cross-cuts. Hence, by § 48, II., it is possible to draw q other cross-cuts Q_2, which likewise do not divide the surface, by means of which a surface T_2 may be formed; the question then arises, whether T_2 must be simply connected. Let us form, as in § 49, a new system of surfaces $\mathfrak{C}$ from T_1 and T_2 in two ways, first by drawing the lines Q_2 in T_1, and secondly by drawing the lines Q_1 in T_2. As in § 49, let the number of cross-cuts which the lines Q_2 form in T_1 be denoted by $q + m$. Then, according to the first part of the proof of the fundamental proposition (§ 49 or § 51), the number of cross-cuts which the lines Q_1 form in T_2 is likewise $q + m$. Now T_1 is, according to the assumption, a single simply connected surface, which is transformed by $q + m$ cross-cuts into the system of surfaces $\mathfrak{C}$; therefore $\mathfrak{C}$ consists of $q + m + 1$ distinct pieces, each by itself simply connected (§ 48, V.). But the same system is also produced from T_2 by $q + m$ cross-cuts; therefore the surface T_2, which consists of one piece, has the property that it is resolved by $q + m$ cross-cuts into $q + m + 1$ distinct pieces, each by itself simply connected. Thus, by § 48, IX., T_2 is in fact simply connected.

Upon this is based a classification of surfaces and the more exact determination of their connection.

If a surface be multiply connected, a cross-cut can be drawn in it which does not divide the surface (§ 48, II.). If the case occur that, after the addition of this first cross-cut, the surface has become simply connected, then according to the last proposition it is changed into a simply connected surface by *every* cross-cut which does not divide it. In this case the surface is said to be *doubly connected*.

But if, after the first cross-cut is drawn, the surface remain multiply connected, a new cross-cut can be drawn which does not divide it. If this change it into a simply connected surface, the same result is obtained by means of any other two cross-cuts which do not divide the surface. The surface is then said to be *triply connected*.

If the surface be still multiply connected after the addition of the second cross-cut, a third can then be drawn which does not divide the surface, and, according as the resolution into a simply connected surface is effected by means of *three*, *four*, etc., cross-cuts, the surface is said to be *quadruply*, *quintuply*, etc., connected.

In general, *a surface is said to be $(q + 1)$-ply connected when it can be changed by means of q cross-cuts into a simply connected surface.* In that case it is unimportant how the cross-cuts are drawn, provided only that none of them divide the surface.

SIMPLY AND MULTIPLY CONNECTED SURFACES. 215

But after the surface becomes simply connected, it is no longer possible to draw a cross-cut in it which does not divide the surface (§ 48, IV.).

53. We are now prepared to prove some propositions, in part relating to the variation or non-variation of the order of connection,[1] in part relating to boundary-lines.

I. *The order of connection of a surface is diminished by unity by every cross-cut which does not divide it.*

For, if the surface be $(q+1)$-ply connected, it follows from the second proposition of § 52 that, however the first cross-cut which does not divide the surface may be drawn, the resolution into a simply connected surface is always effected by means of $q-1$ cross-cuts; hence the new surface is q-ply connected.

II. *If a line be drawn from a point a of the boundary into the interior of the surface, and if, without returning into itself, it end in a point c in the interior of the surface, such a line does not change the order of connection of the surface.* (If a section be made along this line, it is called a *slit.*)

Call the original surface T and that formed by the introduction of the line ac, T'. In the first place it is evident that, if T be simply connected, T' must also be simply connected; for, if every closed line in T form by itself alone the complete boundary of a portion of the surface, so also does every closed line in T', i.e., every line which does not cross ac. Therefore let T be multiply connected, say $(q+1)$-ply. Then a cross-cut which does not divide the surface can always be drawn in T (§ 48, II.), and in fact so that the line ac forms part of the same. This is always possible; for, if the cross-cut be drawn as directed in § 48, II., with the help of a closed line which does not form by itself alone a complete boundary, it can be made to run from a point of the latter on each side to the edge of the surface in an entirely arbitrary way; therefore, since a lies on the edge, so that ac always forms part of the same. Let this cross-cut be denoted by acb, and let the surface formed from T by means of it be called T''; then the latter

[1] Sometimes called *connectivity*. (Tr.)

surface is q-ply connected (I.). But T''' can also be formed from T' by means of the line cb, and this line forms in T'' a cross-cut which does not divide the surface, because it is so drawn that the contiguous portions of the surface on both sides of it are connected. Consequently T'' has the property, that it is changed into a q-ply connected surface by means of one cross-cut which does not divide the surface; therefore T'' is $(q+1)$-ply connected, just as T was.

NOTE.— This proposition remains perfectly valid, if the internal point c be a branch-point.

III. *If a single point c be removed from a surface T at any place, the order of connection is thereby increased by unity.*

Let the surface formed by the removal of the point c be called T'. Connect the point c with any point a of the boundary of T^1 by means of a line which does not intersect itself, thereby forming a new surface T''. Then the latter can also be regarded as a surface formed from T by drawing the line ac, which starts from a boundary-point a and ends in an interior point c; therefore the order of connection of T'' is the same as that of T (II.). On the other hand, ac is a cross-cut in T' which does not divide the surface, since we can pass round c from the one side to the other. Accordingly, by I., T' is of an order higher by unity than T'', and therefore also than T.

NOTE.— The preceding does not lose its validity if the point removed be a branch-point.

IV. *If an (actually) closed line, forming by itself alone the complete boundary of a portion of a surface, be drawn in any position in a portion which contains either no branch-point or at most one (of any order [§ 13]), and if the portion bounded by this line be removed from the surface, the order of connection is thereby increased by unity.*

For the order of connection will not be changed, if the boundary-line which bounds the piece removed be more and

[1] If T be a closed surface, it is assumed that it already possesses a boundary-point a (§ 46).

more contracted. But if it finally shrink into a point, the case is the same as the preceding. Hence this proposition holds, if the piece removed contain either no branch-point or only one. But if it contain more than one, it would no longer be possible to let the boundary-line shrink into a point.

Cor. In the case of a surface closed at infinity, it is necessary to assume in some place a boundary-point (§ 46). This may itself also be a branch-point. *If a piece which contains this boundary-point, but no other branch-point, be removed from such a surface, the order of connection is not changed.* For we can let the boundary-line which bounds the piece removed contract into the boundary-point, and thus obtain again the original surface.

V. *If a $(q+1)$-ply connected surface T be resolved by a cross-cut R into two distinct pieces A and B, each of the latter has a finite order of connection; and if r and s be the numbers of cross-cuts which determine these orders, then $r + s = q$.*

The surface T is reduced to two simply connected pieces by means of $q+1$ cross-cuts, of which the first q do not divide the surface. If, however, the dividing cross-cut R be first drawn, we cannot immediately infer the truth of the above enunciation from Riemann's fundamental proposition, because that proposition already assumes what will here be first proved, namely, that now also after the addition of a finite number of cross-cuts simply connected pieces are finally obtained again. We remark that all the cross-cuts running in A can be so placed that they meet the cross-cut R either not at all or only in one of its end-points. For, since R, except for its end-points, lies entirely in the interior of the surface, and since therefore a zone free of boundary-lines exists on each side of it, we can displace along the line R the end-points of all the cross-cuts which meet R until they coincide with an end-point of the same. But then every cross-cut which does not divide A is also a non-dividing cross-cut in T. From this it follows that the number r of non-dividing cross-cuts possible in A cannot be greater than q; for, otherwise, it would also be possible to draw in T more than q non-dividing cross-cuts, and this is contrary to the hypothesis that this surface is $(q+1)$-ply connected (§ 52). Therefore r is a finite number, and when these r cross-cuts have been drawn in A, another non-dividing cross-cut is no longer possible; thus A has become simply connected (§ 48, VI.). Exactly the same holds for B. Here, too, all cross-cuts can be so drawn that they likewise form cross-cuts in T, and if s be the number of non-dividing cross-cuts possible in B, s cannot be greater than q and is therefore a finite number. By drawing these s cross-cuts B is made simply connected. Now if all these cross-cuts, and R also, be drawn, we have obtained two simply connected pieces by means of $r+s+1$ cross-cuts. Therefore, according to Riemann's fundamental proposition,

$$r + s = q.$$

Of these q cross-cuts, r run entirely in A, the other s entirely in B.

VI. *If a $(q+1)$-ply connected surface T be resolved by means of ν non-dividing and s dividing cross-cuts (which may be drawn in any order) into $s+1$ distinct pieces $A_0, A_1, A_2, \cdots, A_s$, and if the orders of connection of these pieces be determined by the numbers of cross-cuts $r_0, r_1, r_2, \cdots, r_s$, respectively, then*

$$q = \nu + r_0 + r_1 + r_2 + \cdots + r_s.$$

Since by the last proposition a surface of finite order is always resolved by a dividing cross-cut into two pieces which are also of finite orders, these orders will still be finite if a series of non-dividing cross-cuts be drawn in T before the division. The same conclusion holds, if each of the resulting pieces be further resolved in like manner. Therefore $r_0, r_1, r_2, \cdots, r_s$ are finite numbers, none of which is greater than q, in whatever order we may have drawn the $\nu + s$ cross-cuts. Now if all the pieces A be further changed into simply connected surfaces by means of their respective r cross-cuts, we have finally $s+1$ simply connected pieces, which are formed by means of $\nu + s + r_0 + r_1 + r_2 + \cdots + r_s$ cross-cuts in all. But we likewise obtain $s+1$ simply connected pieces, if we first

make T simply connected by means of q cross-cuts and then by means of s additional cross-cuts resolve it into $s+1$ pieces. Therefore,
$$q+s = \nu+s+r_0+r_1+r_2+\cdots+r_s,$$
or
$$q = \nu+r_0+r_1+r_2+\cdots+r_s.$$

VII. *If a $(q+1)$-ply connected surface be resolved by m cross-cuts into two distinct pieces, one of which, S, is simply connected, then the other, T', is $(q-m+2)$-ply connected, i.e., only $q-(m-1)$ cross-cuts are needed to modify it into a simply connected surface.*

Let x be the number of cross-cuts which change T' into a simply connected surface T_0'. If these cuts be drawn, we have two simply connected surfaces, T_0' and S, formed by means of $m+x$ cross-cuts. But if the original surface be first modified into one simply connected by means of the appropriate q cross-cuts, and if the surface so formed be then divided into two distinct pieces by means of an additional cross-cut, we have again formed two simply connected surfaces by means of $q+1$ cross-cuts. Then, by the fundamental proposition (§ 49),
$$(m+x)-2 = (q+1)-2,$$
and hence
$$x = q-(m-1).$$

VIII. *If a surface consisting of one piece possess more than one boundary-line, i.e., if its boundary consist of several distinct closed lines, it is multiply connected.*

If a and b be two points situated on different boundary-lines, we can, since the surface is connected, draw a line from a to b through the interior of the surface. This is a cross-cut, which does not divide the surface, however, for we can come from one side of the cross-cut to the other side of the same in the surface by following one of the two boundary-lines. Since it is thus possible to draw in the surface a cross-cut which does not divide it, the surface is multiply connected (§ 48, I.).

IX. From this follows: *A simply connected surface always possesses a single boundary-line, i.e., its boundary can be traced in a continuous description.* (*Or, its boundary consists of only a single point.*)

Therefore, after a multiply connected surface has been modified into a simply connected surface by means of cross-cuts, wherein the cross-cuts, as they are drawn, are to be added to the original boundary as new boundary-pieces, it must be possible to trace the latter, together with the original boundary, in a continuous description. Therein each cross-cut simultaneously forms the boundary of each piece of the surface contiguous to it on each side. If then the entire boundary be traced in the positive direction, so that the bounded region always lies on the left of the boundary, each cross-cut must be traced twice in opposite directions. (Cf. Figs. 43 to 46, pp. 191 and 192.)

X. *The number of boundary-lines is either increased or diminished by unity by every cross-cut.*

According to the discussion of § 47 a cross-cut always furnishes two boundary-edges, because it simultaneously bounds the portions of the surface contiguous to it on each side. Now there are three kinds of cross-cuts (§ 47):

(1) The cross-cut may join two points a and b of the same boundary-line. The latter is then divided into two parts by the points a and b; one part forms with the one edge of the cross-cut one boundary-line, the other part forms with the other edge a second boundary-line. Thus two boundary-lines are formed from one, and the number of boundary-lines is increased by unity.[1]

(2) The cross-cut may join two points which lie on different boundary-lines. Then it unites these into a single one, because its two edges establish the connection. Thus from two boundary-lines is formed one, and the number of boundary-lines is diminished by unity.

(3) The cross-cut may begin at a point of the boundary and

---

[1] This result still holds if the points a and b approach each other and finally coincide.

end at a point of its previous course. Then one of its edges, together with the boundary-line from which it starts, forms a single closed boundary-line. But in addition the inner edge of its closed portion forms a new boundary-line, so that the number of boundary-lines is increased by unity.

XI. *If a closed surface (which therefore possesses but one boundary-point) be multiply connected, and if it can be modified into a simply connected surface by means of a finite number of cross-cuts, the number of such cross-cuts is always even.*

Let the given surface be $(q+1)$-ply connected, so that q cross-cuts modify it into a simply connected surface. Since the surface originally possesses only a single boundary-point, the number of its boundary-lines is 1. This number is either increased or diminished by unity by every cross-cut (X.). Let p be the number of cross-cuts which produce an increase, and therefore $q-p$ the number which produce a diminution in the number of boundary-lines; then at the end the number of boundary-lines is $1+p-(q-p)$. But since the surface is then simply connected, it possesses only one boundary-line (IX.); accordingly we have

$$1+p-q+p=1,$$

and hence $\qquad q=2p.$

Thus q is an even number.

54. If we know the number of sheets, as well as branch-points, in a surface closed at infinity, we can determine its order of connection. For this purpose, as in § 13, we regard a winding-point of the $(m-1)$th order as resulting from the coincidence of $m-1$ simple branch-points. If in this sense g be the number of simple branch-points, n the number of sheets, and q the number of cross-cuts which modify the surface into one simply connected, we can find a relation between these three numbers.[1]

[1] Cf. with the following: Roch, "Ueber Funktionen complexer Grössen," Schlömilch's *Zeitschr. f. Math.*, Bd. 10, S. 177.

Let A_0 be the boundary-point to be assumed in the closed surface. We will also remove from the surface $n-1$ other points $A_1, A_2, \cdots, A_{n-1}$, one from each of the $n-1$ other sheets; for greater simplicity let us assume that these n points lie directly below one another. Now, since the order of connection is increased by unity through the removal of every such point (§ 53, III.), the order in this case is increased by $n-1$. Therefore $q+n-1$ cross-cuts are necessary to modify the surface into a simply connected surface after the removal of the $n-1$ points $A_1, A_2, \cdots, A_{n-1}$. In this $(q+n)$-ply connected surface let us now draw cross-cuts in the following way. From each point A let us draw lines to all the branch-points which lie in the same sheet with A. Then it is evident that thereby we have actually formed cross-cuts, because we can pass through a branch-point into all those sheets which are connected at that point. If two points A_ι and A_k lie in two sheets which are connected at a simple branch-point a, then $A_\iota a$ and aA_k together form a line which leads from one boundary-point A_ι through the interior of the surface to a boundary-point A_k; thus $A_\iota a A_k$ is a cross-cut. On the other hand, if a be a winding-point of the $(m-1)$th order, at which are connected m sheets containing the points $A_1, A_2, \cdots, A_m$, say, then first $A_1 a A_2$ is a cross-cut, but after it is drawn each one of the $m-2$ other lines $aA_3, aA_4, \cdots, aA_m$ becomes a cross-cut; therefore in this case we have in all $m-1$ cross-cuts, just as many as there are simple branch-points united in a. If we proceed in this way with all the branch-points, we obtain exactly as many cross-cuts as there are simple branch-points, i.e., g. But these g cross-cuts resolve the surface into n distinct pieces, each by itself simply connected; that is, the n sheets of the surface are separated from one another by them in a certain manner. For, if p_ι and p_k be two points lying one above the other in any two sheets, we can come from p_ι to p_k only by crossing branch-cuts and winding round branch-points; but the latter is rendered impossible by the cross-cuts constructed. Thus every two such points always lie in distinct pieces. Only the points A furnish exceptions. We can always

SIMPLY AND MULTIPLY CONNECTED SURFACES. 223

come from any one point A to another point A by crossing a branch-cut. The n sheets of the surface are therefore separated from one another in this way: in every sheet an angular piece (or also several such pieces), which is formed by two cross-cuts meeting at the point A, is separated from the sheet, and in its place enters a corresponding angular piece of another sheet. Accordingly the surface consists of n distinct portions. But each of these is by itself a connected piece; for, since it is closed at infinity, its boundary consists solely of the cross-cuts which meet in the point A. For the same reason, also, each portion is by itself simply connected, since we can come from only one side of every closed line drawn in it to that boundary; thus each closed line forms a complete boundary. Therefore, after the removal of the $n-1$ boundary-points A_1, A_2, $\cdots$, A_{n-1}, the given surface is resolved by g cross-cuts into n distinct pieces, each by itself simply connected. But this surface was $(q+n)$-ply connected; therefore $q+n-1$ cross-cuts are necessary to modify it into a simply connected surface. Now, to divide the latter into n distinct pieces, $n-1$ additional cross-cuts are necessary (§ 48, V.); accordingly the original surface is divided into n distinct simply connected pieces by means of $q+2(n-1)$ cross-cuts. But the same number was found above equal to g, and therefore, by the fundamental proposition (§ 49),

$$g = q + 2(n-1) \text{ or } q = g - 2(n-1).$$

Compare with this result the examples given in § 46 to which it is applicable. In the third, $n=2$, $g=2$; therefore $q=0$, and the surface is simply connected. In the fifth example $n=2$, $g=4$; therefore $q=2$, and the surface is triply connected.

From the result obtained above we may draw certain conclusions. For, since in a closed surface q is always an even number (§ 53, XI.), g must also be even. Therefore a surface closed at infinity always possesses an even number of simple branch-points. With an n-sheeted surface the simplest case would be the occurrence of two branch-points of the $(n-1)$th order; and if this be the case, the surface is simply connected.

A further inference, which follows from the preceding, is this, that a surface, which serves to distribute the values of an *algebraic* function w so that this becomes a uniform function of position in the surface (§ 12), always possesses a finite order of connection, and therefore can be changed by a finite number of cross-cuts into a simply connected surface. For the number of sheets n is equal to the number of values which the function w possesses for each value of the variable, and is therefore a finite number. That the number g of simple branch-points is also finite, follows from the fact that the branch-points are to be sought for only among those points at which values of the function become either equal or infinite (§ 8). The number of the latter points is finite (§ 38). But if $F(w, z) = 0$ denote the equation of the nth degree by which w is defined, the points z at which values of the function become equal are those which simultaneously satisfy the equations

$$F(w, z) = 0 \text{ and } \frac{\delta F(w, z)}{\delta w} = 0.$$

By the elimination of w from these equations we obtain an equation of finite degree in z. Moreover, since at most n values can become equal at each of these points, and since therefore at each branch-point n sheets at most can be connected, each branch-point is of finite order. Accordingly n and g are finite numbers, and hence q is also finite.

55. From the result of the preceding paragraph we can also derive a relation between the order of connection of an unclosed surface extended in a plane, the number of its simple branch-points and the number of circuits made by its boundary.

We begin with a surface closed at infinity. Let this be $(q'+1)$-ply connected; let g' be the number of its simple branch-points and n the number of its sheets. Then, according to the preceding paragraph, we have

$$q' = g' - 2(n - 1).$$

We will now assume that the boundary-point which is to be assigned to the surface lies at the point at infinity of one of its sheets, and first premise that in no sheet is the point at infinity a branch-point. If we then remove from each sheet a piece which contains its point at infinity, and which is therefore bounded by a line returning simply into itself, the order of connection of the surface is increased by unity for every piece removed, with the exception of that which contains the assumed boundary-point (§ 53, IV.). Thus the order of connection is increased by $n-1$. If, therefore, the new surface be $(q+1)$-ply connected, we have

$$q = q' + n - 1,$$

and consequently $\quad q = g' - n + 1.$

But after the points at infinity have been removed from the surface, we can assume its sheets, which were previously to be conceived as infinitely great spherical surfaces, to be again extended in the plane. Each sheet then appears bounded by a line returning simply into itself, which makes a positive circuit if it be described in the direction of increasing angles. Consequently, if U denote the number of circuits forming the boundary, we have $U = n$. The number of simple branch-points g contained in the new surface is equal to the previous number g', since by the assumption no branch-point was removed from the surface. We therefore obtain from the last equation

$$q = g - U + 1.$$

This is the relation mentioned above, and it will now be shown that it does not lose its validity when certain changes are made in the surface.

Let us first consider the case when m sheets in the original surface are connected at a point at infinity, when therefore $m-1$ simple branch-points are united in that point. Then the number of pieces removed is no longer equal to n as before, but since one of them is bounded by a line which makes m circuits round a branch-point, and since it thus takes the place of m of the previous pieces, that number is equal to only $n - m + 1$. Moreover, that piece which contains the assumed boundary-point does not cause any increase in the order of connection; accordingly that increase amounts to $n - m$, or

that is,
$$q = q' + n - m,$$
$$q = g' - 2(n - 1) + n - m$$
$$= g' - m + 1 - n + 1.$$

When the sheets are extended in the plane the number of circuits U is again equal to n; for the only change in this respect is that all the n boundary-lines are no longer distinct, but m of them are united into a single one, which now, however, makes m circuits. On the other hand, $m - 1$ simple branch-points are removed from the surface with the points at infinity; therefore now
$$g = g' - m + 1.$$
If we substitute this value of g' in the last equation, we obtain again as before
$$q = g - U + 1.$$

We will now modify the n-sheeted surface extended in the plane by removing places in the interior.

Let us first consider a closed line bounding a portion of the surface which does not contain a branch-point, and let us imagine this piece removed. Then, in the first place, q is increased by $+ 1$ (§ 53, IV.). But the new boundary-line, if its boundary-direction is to be positive, must be described in the direction of decreasing angles. Therefore, if we now understand in general by the *number of positive circuits* the positive or negative number U, which results from subtracting the number of circuits in the direction of decreasing angles from the number of circuits in the direction of increasing angles, this number U in the preceding case must be increased by $- 1$. At the same time q is increased by $+ 1$, and thus the above relation remains unchanged.

For instance, if the surface consist of one sheet, then $g = 0$; if, further, it be bounded by one outer line and k smaller circles enclosed by the former, then the outer line makes a circuit in the direction of increasing angles for a positive boundary-direction, but each of the inner circles a circuit in the opposite direction; therefore

$$U = 1 - k,$$
and we obtain $\quad q = k - 1 + 1 = k;$

thus the number of cross-cuts is equal to the number of inner circles.

Secondly, if a piece of the surface be removed which contains a branch-point of the $(m-1)$th order, the boundary-line of which therefore makes m circuits, q is again increased by $+1$ (§ 53, IV.), U at the same time by $-m$, g by $-(m-1)$, and therefore $g - U$ by $+1$; consequently the above relation again remains the same.

After the modifications introduced hitherto the surface considered has a configuration such that the outer boundary-lines enclose all branch-points situated in the finite part of the surface; it also has gaps in the interior, yet of such a kind that each of the pieces of surface removed contains either no branch-point or only one (of any order). We have now to inquire whether the above relation changes, either when the outer boundary-lines no longer enclose *all* finite branch-points, or when the inner boundary-lines enclose portions of the surface that were removed in which more than one branch-point was contained. Both conditions lead to the same inquiry; namely, to the examination of the case when there is removed a portion of the surface contiguous to an (outer or inner) edge which contains a branch-point of the $(m-1)$th order, but no gaps. The latter assumption can be made without loss of generality, since the occurrence of gaps has already been disposed of by the preceding considerations. Now if such a piece of the surface is to be removed, then, since it is contiguous to the edge, its removal must be effected by means of cross-cuts, and these must be drawn in such a way that the boundary of the piece removed, consisting of the cross-cuts and the contiguous parts of the boundary, forms a closed line which makes m circuits round the winding-point. If none of these cross-cuts wind round the branch-point, then m cross-cuts are necessary to that end; otherwise a less number. Since, however, the piece removed is bounded by a single actually closed line, it is simply connected (§ 46, Ex. 2). We will now examine the case in which no cross-cut winds round the branch-point; then a simply connected piece of the surface is removed by means of m cross-cuts, and consequently the order of connection of the surface which remains is diminished by $m-1$ (§ 53, VII.). At the same time g is diminished by $m-1$ through the removal of the winding-point of the $(m-1)$th order. But the number U suffers no change. For, since the m cross-cuts add no new positive or negative circuits, merely a different kind of connection of the boundary-line is produced by the removal of the branch-point, while the circuits of the same remain unchanged. Thus, since q and g are each diminished by $m-1$ and U remains unchanged, the above relation still holds.

Finally, let us consider the case in which the boundary is changed by means of cross-cuts which do not divide the surface. In this we turn our attention to the change of direction which the lines experience, and remark that a line makes a positive circuit if it experience a total change of direction equal to 2π. If a non-dividing cross-cut be drawn in the surface, this at the same time furnishes two boundary-pieces and must be described twice in opposite senses in conforming to the positive boundary-direction. Where the cross-cut meets a part of the boundary of the surface, the boundary-direction experiences an abrupt change. Let a be the angle by which the direction changes (Fig. 50). (The case can, it is true, occur in which the cross-cut changes into a part of the boundary without an abrupt change of direction, but this is included in the preceding if we assume $a = 0$.) In the description of the boundary-lines, of which the cross-cut always forms a part, we once more return to the

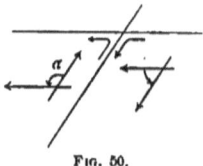

Fig. 50.

former place, since the cross-cut must be described twice; then the cross-cut is described in the opposite direction, but the contiguous part of the original boundary in the same direction as before. From this it follows that the boundary-direction now experiences an abrupt change equal to $\pi - \alpha$. Consequently the end-point of the cross-cut produces a total change of direction equal to π. (Also in the case when this ends in a point of its previous course, because then the cross-cut itself takes the place of the original boundary-line.) The same change occurs at the other end of the cross-cut. Therefore the cross-cut causes at its end-points a change of direction equal to 2π. On the other hand, the change of direction experienced by the cross-cut during its course need not be considered, because this change is neutralized by that of the subsequent description in the opposite sense. Consequently each non-dividing cross-cut increases the number of positive circuits U by $+1$;[1] at the same time, however, it diminishes the order of connection by unity (§ 53, I.) and therefore q is increased by -1. Consequently the above relation holds in this case also.

According to the preceding considerations, the equation $q = g - U + 1$ holds for all surfaces which can be formed by

[1] The same conclusion holds when a cross-cut divides the surface into two distinct pieces. For, since a cross-cut which joins two different boundary-lines never divides the surface (§ 53, VIII.), a dividing cross-cut can either merely join two points of the same boundary-line, or, starting from one boundary-line, end in a point of its previous course. In both cases two boundary-lines are produced by it from one. (§ 53, X. (1) and (3). See also Fig. 40 and Fig. 41, p. 190.) If these be traced in succession in the positive boundary-direction, the original boundary-line is described once, but the cross-cut twice in opposite directions. Consequently the above considerations still hold. Therefore if U be the number of circuits of the original boundary-lines, U_1 and U_2 the numbers for the two boundary-lines resulting from the cross-cut, we have
$$U_1 + U_2 = U + 1.$$
It is evident at once that this formula does not lose its validity, if the two boundary-lines resulting from the cross-cut have only one point in common, in which case two pieces of the original boundary-line approach each other and the cross-cut drawn at this place is infinitesimal in length.

230   THEORY OF FUNCTIONS.

means of the cross-cuts discussed. For the formation of many kinds of surfaces (as, for instance, those represented in Riemann's dissertation: "Lehrsätze aus der Analysis situs," u. s. w., *Crelle's Journ.*, Bd. 54, S. 110, last example), it would be necessary still to consider the case in which a portion of the surface contiguous to an edge and containing a branch-point is to be removed by means of cross-cuts which wind round the branch-point; in another place[1] it was shown that in this case also the above relation does not lose its validity. But we cannot always affirm with certainty that every surface, however bounded, could be produced by means of such cross-cuts, as long as we do not know in advance the form of a particular given surface. Therefore we will in preference add another proof for the general validity of the above relation.

This is based upon the property, that the boundary-line of a simply connected surface, extended in the plane and containing no branch-points, always makes but one circuit, and that this also holds when the surface has first been reduced to this simple connection by means of cross-cuts. For in the first place only one boundary-line can ever occur in such a surface (§ 53, IX.).

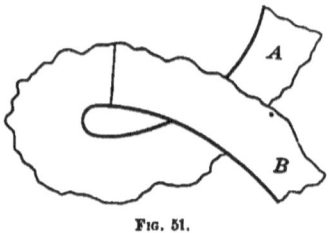

Fig. 51.

But if we represent this as a movable thread, we can show that it can always be deformed into a circle which is to be described once. For, since the boundary-line nowhere intersects itself, and also since its displacement is nowhere prevented by a branch-point, the deformation into a circle could be made impossible only by the line somewhere forming a loop which could not be opened by enlarging. But if this be the case, the portions of the surface which are contiguous to the boundary-line where this forms the loop, and

[1] "Zur Analysis situs Riemann'scher Flächen," *Ber. d. Wien. Akad.*, Bd. 69, Abth. II., Januar 1874. See here Fig. 1.

SIMPLY AND MULTIPLY CONNECTED SURFACES. 231

which thus pass one over the other, must later be connected with each other in their continuations beyond A and B (Fig. 51). If these portions always remain separated beyond A and B, the loop can be at once opened by enlarging. But if A and B be connected, we can, by drawing a cross-cut from a point of the loop, come from the one side of the same to the other beyond A and B, since these are connected; thus the cross-cut does not divide the surface and this is multiply connected (§ 48, I.). Accordingly every loop which occurs in a simply connected surface can always be opened, and the boundary-line therefore be deformed into a circle. If the latter be described in the direction of increasing angles, it forms a single positive circuit.

Assume now an arbitrary Riemann surface extended over a finite part of the plane, and let q, g, U have for this their former meanings. Then $g - U + 1$ is always an integer (or zero). The formula to be proved asserts that this number is exactly equal to q. We will now not presuppose this, but will assume

$$g - U + 1 = q + k,$$

and then prove that k must be zero. To this end we first remove all branch-points from the surface, by enclosing each one in an actually closed line and removing from the surface the piece so bounded which contains the branch-point. Then the last equation still holds; for, as was previously shown, p. 227, for the removal of a branch-point of the $(m-1)$th order, q changes into $q + 1$, g into $g - m + 1$, U into $U - m$, and hence $g - U$ into $g - U + 1$. Therefore, if q change into q', U into U' after the removal of all the branch-points, we obtain, since g becomes zero,

$$-U' + 1 = q' + k.$$

For the modification of the surface into one simply connected, q' non-dividing cross-cuts are requisite. If these be drawn, q' changes into $q' - 1$ for each one, and, by p. 229, U' at the same time into $U' + 1$. Consequently the preceding equation still holds. Hence if U' change into U'' when the surface becomes

simply connected, and when therefore q' becomes zero, we have
$$-U'' + 1 = k.$$
But now this surface is not only simply connected, but it no longer contains a branch-point; its boundary-line therefore makes but one circuit, *i.e.*, $U'' = +1$, and consequently $k = 0$; which was to be proved.

For a simply connected surface ($q = 0$) the equation
$$q = g - U + 1$$
changes into $\qquad U = g + 1.$

Accordingly the proposition, which we found to be valid for a special case, § 13, holds generally: The number of circuits of the boundary-line of a simply connected surface is greater by unity than the number of simple branch-points in its interior. Yet it is well to notice that the validity of this relation, as well as of the more general one $q = g - U + 1$, depends upon the surface being extended in the plane.

56. We will now also examine such Riemann surfaces as cannot be extended in a plane, inquire under what conditions their orders of connection are finite, and determine these orders more exactly. At the same time we first premise that the surface possesses only a finite number of sheets and branch-points, and assume that none of its boundary-lines pass through a branch-point.

We begin with a closed surface, which therefore possesses only one boundary-point a. We will call this a *complete surface* and designate it by W. By § 54, the relation
$$Q = G - 2(n-1) \qquad (1)$$
holds for this surface, if G denote the number of simple branch-points contained in it, n the number of its sheets, and Q the number of its cross-cuts. Accordingly, Q is a finite number, if G and n be finite. Hence our further investigations relate exclusively to the boundary-lines.

The boundary-lines which are to be introduced into a complete surface must be furnished by cuts drawn in the surface.

Either these do not divide the complete surface, or they divide it and remove from it single surface-pieces. Accordingly we distinguish two kinds of boundary-lines.

By *boundary-lines of the first kind* we understand such as do not remove a piece from a complete surface. They are therefore characterized by the condition that in the new surface T two edges, which belong either to different boundary-lines or to one and the same boundary-line, run everywhere infinitely near each other. If we consider only the lines along which the cuts are made, without regarding the edges furnished by them, these lines form a line-system, and the portions of the surface T itself which are contiguous to the two sides of each line belong with that line.

Secondly, it may happen that, when pieces are removed from W by the cuts introduced, so that gaps occur in T, the edges belonging to the boundary-lines which are furnished by the cuts, likewise run for considerable distances infinitely near one another in isolated places. We will, however, lay particular stress on those boundary-lines in connection with which this is not the case, and call them by the distinctive name *boundary-lines of the second kind*. Consequently a boundary-line of the second kind is formed by a closed line, and everywhere on one side of this line there borders a portion of the surface belonging to T, while on the other side a gap occurs. Thus in connection with a boundary-line of the second kind two edges never run infinitely near each other for any distance;[1] on the contrary, when in connection with boundary-lines which divide the surface two edges do so run, a boundary-line of the first kind is connected with a boundary-line of the second kind. If the boundary-lines of the second kind be conceived in this way, each boundary-line is either of the first or of the second kind, or it is a combination of the two.

We will now show that every surface T, which possesses arbitrary boundary-lines, can be obtained from an appropriate

[1] The case in which edges, which belong to boundary-lines of the second kind, come infinitely near one another at isolated points will be considered later.

complete surface by means of cuts. Let us first assume that T contains only boundary-lines of the second kind. In this case it can first be made into a complete surface W by the addition of surface-pieces B, and then be obtained again from this surface by means of cuts. This is at once evident, if each one of the boundary-lines which enclose the gaps run entirely in one and the same sheet. If, on the other hand, a boundary-line run in several sheets, we assume that the supplementary piece B consists of the same sheets, by imagining each sheet extended beyond the edge. In a place where the boundary-line passes from one sheet into another, a branch-cut must occur, or be capable of being assumed, in T. At such a place we extend the branch-cut into B, let it end in B at a branch-point, and assume that the connection of the sheets at this point in B is just as it actually occurs in T. This can be effected in every place where it is necessary independently of every other place, and depends only upon the particular connection of the sheets in T for each branch-cut. If a gap be bounded by several boundary-lines, the same method of procedure is followed for each. We thereby obtain a surface B everywhere contiguous to the boundary-lines, which contains no gaps, and which is also completely bounded by these boundary-lines, since the latter bound completely the gaps.

If any surface T, which possesses arbitrary boundary-lines, be under discussion, we imagine all boundary-lines of the first kind removed, by regarding the lines along which they run as not drawn. We then supplement the surface, in the manner outlined above, into a complete surface and cut from this first the boundary-lines of the second kind. This done, it is at once evident that the boundary-lines of the first kind, whether they occur alone or in connection with those of the second kind, can be cut in the surface.

Thus we can always regard a given surface T as one formed from a complete surface W by means of cuts. Let the number of surface-pieces B removed from W by boundary-lines of the second kind be s. All cuts made in W run along certain lines. We will now suppose these lines to be drawn in W; then they all together form a line-system. It is not necessary for all the lines of this system to be connected. We will assume that it consists of r distinct systems $L_1, L_2, \ldots, L_r$. Each L forms a connected line-system complete in itself, but it may at the same time contain several boundary-lines. Let us assume the boundary-point a, which is to be assigned to W, on one of the systems L, and let a boundary-point be also taken on each of the $r-1$ other systems. Then all the systems are exactly alike in this respect,—that each one is connected with a boundary-point. We therefore need to examine more closely only one of these systems; we will designate it indefinitely by L, and the boundary-point on it by a.

Since L is by itself wholly connected, and hence, with an exception to be mentioned immediately, contains no simply closed lines, it can be decomposed into simple line-segments (§ 50). (The exception referred to occurs when L consists of a single simply closed line, which therefore begins at a point a and also ends at that point; but in that case L forms *one* cross-cut.) If we denote, as in § 50, the numbers of end- and nodal-points contained in L by $e, k_3, k_4, k_5, \ldots$, then, by § 50, L consists of

$$\tfrac{1}{2}(e + k_3 + 2k_4 + 3k_5 + \cdots)$$

simple line-segments; or of

$$\tfrac{1}{2}(e + K)$$

simple line-segments, if for brevity we let

$$k_3 + 2k_4 + 3k_5 + \cdots = K.$$

A simple line-segment is a cross-cut, if both of its end-points lie on an edge; but if one end-point lie in the interior of the surface, the simple line-segment is a slit (§ 53, II.). Hence the system L in general consists partly of cross-cuts, partly of slits. But the latter need not be considered, because a slit, which begins at the edge and ends in the interior, can never divide the surface and does not change the order of connection (§ 53, II.); its effect, on the contrary, consists only in extending a boundary-line which already exists. Hence it is important to determine the number p of cross-cuts contained in L. To that end we remark that the line-system L, according to Lippich's proposition, can in fact be regarded as a system of cross-cuts, but only when conditions (1) and (2) of § 51 are fulfilled. This is not the case if the end-points of L lie in the interior of the surface. We can, however, remove these by cutting off every line-segment which contains such a point at the nodal-point nearest to it. Then the system which remains, since it satisfies conditions (1) and (2), consists of only a definite number p of cross-cuts; but the line-segments cut off become slits. We will now always place the boundary-point a, which can indeed be arbitrarily assumed, either at an end-point or at an ordinary point. If a be an end-point, only $e-1$ line-segments are to be cut off, since a, as a boundary-point, need not be removed. To find p, therefore, we must deduct from the number $\frac{1}{2}(e+K)$ of all the simple line-segments the number $e-1$ of segments which are not cross-cuts, and we thereby obtain

$$p = \tfrac{1}{2}(e+K)-(e-1) = \tfrac{1}{2}(K-e)+1.$$

We obtain the same value if a be an ordinary point. Then all the e end-points must be removed; but now, in order to decompose the system which remains into cross-cuts, since all the line-segments which end at a must be regarded as possessing end-points at that place, we must, according to the discussion of § 51, count a as two end-points in addition. Thus the whole number of simple line-segments is now

$$\tfrac{1}{2}(2+e+K),$$

and therefore

$$p = \tfrac{1}{2}(2+e+K) - e = \tfrac{1}{2}(K-e)+1,$$

as before; and this relation is also valid for the exceptional case mentioned above.

Consequently each of the r line-systems L contains

$$\tfrac{1}{2}(K-e)+1$$

SIMPLY AND MULTIPLY CONNECTED SURFACES. 237

cross-cuts, and therefore they all in the aggregate contain

$$\tfrac{1}{2}(\Sigma K - \Sigma e) + r$$

cross-cuts.

If we now refer the letters e, k_3, k_4, k_5, $\cdots$, and K to the end- and nodal-points of the entire line-system formed by all the boundary-lines; that is, if we write e and K instead of Σe and ΣK, we obtain for the number p of cross-cuts which are formed by all the boundary-lines the value

$$p = \tfrac{1}{2}(K - e) + r. \qquad (2)$$

Some of these cross-cuts divide the surface; others do not divide it. According to the assumption s surface-pieces B were removed from W; thus, including the piece T which remains, W is divided into $s + 1$ pieces. Hence s is the number of dividing cross-cuts, because no cross-cut can divide a surface into more than two pieces, and none can cross another. (Cf. § 48, V.) If, moreover, we denote by ν the number of non-dividing cross-cuts, we have

$$\nu + s = \tfrac{1}{2}(K - e) + r. \qquad (3)$$

Now by (1) the relation

$$Q = G - 2(n - 1) \qquad (4)$$

held for the surface W; but $r - 1$ boundary-points were removed from W, and therefore Q must be increased by $r - 1$. This $(Q + r)$-ply connected surface has now been divided by $\nu + s$ cross-cuts into $s + 1$ pieces; accordingly the proposition proved in § 53, VI. can be applied. If we denote by q the number of cross-cuts for T, and by q_1', q_2', $\cdots$, q_s' the numbers for the s surface-pieces B which were removed, then by VI., § 53, we have

$$Q + r - 1 = \nu + q + \Sigma q_h'. \qquad (5)$$

From this equation and (3) we obtain

$$q = Q - 1 - \tfrac{1}{2}(K - e) - \Sigma q_h' + s.$$

We can find from this formula when q is a finite number. For, since Q, and by VI., § 53, also every q_λ', is finite, q remains finite if s, K, e be finite, *i.e.*, *if the boundary-lines form a finite line-system* (in the sense of § 50).

If T possess only boundary-lines of the first kind, we can quite generally determine the number of its cross-cuts. For in this case the cross-cuts contained in the boundary-lines are all non-dividing, and therefore Q is first to be increased by $r-1$, and then to be diminished by the number p of cross-cuts which are contained in the boundary-lines. But the slits, which can, moreover, only enter as boundary-lines of the first kind, or as parts of such lines, do not change the order of the surface. We therefore have

$$q = Q + r - 1 - p,$$

or by (4) $\qquad q = G - 2n + 1 + r - p;$

and finally by (2) $\quad q = g - 2n + 1 - \tfrac{1}{2}(K - e), \qquad (6)$

if at the same time the number of simple branch-points contained in T, which in this case is equal to G, be again denoted by g.

But if T also possess boundary-lines of the second kind, we will, in order to obtain a definite expression for q, make a limiting hypothesis; namely, that all the surface-pieces B which are removed can be extended in planes.[1]

For this surface the relation

$$q = g - U + 1$$

of § 55 can be applied. If we denote the numbers of simple branch-points contained in the surface-pieces B by g_1', g_2', $\cdots$, g_s', we have

$$G = g + \Sigma g_\lambda'. \qquad (7)$$

[1] If a complete surface, say a *Riemann* many-sheeted spherical surface, be resolved by any cuts whatever into two distinct pieces, it is quite evident that the cases may occur in which either both pieces or only one of the two can be extended in a plane; but it is very probable that the third case may also occur in which neither can be extended. In the latter case the following investigation would lose its validity.

If, further, U_λ be the number of circuits of the boundary-lines in one of the pieces B, then for this piece

$$q_\lambda' = g_\lambda' - U_\lambda + 1.$$

Therefore, if we let $\quad V = \Sigma U_\lambda,$

we obtain for the aggregate of surface-pieces B

$$\Sigma q_\lambda' = \Sigma g_\lambda' - V + s.$$

If we subtract this equation from (4) and attend to (7), we obtain

$$Q - \Sigma q_\lambda' = g - 2n + 2 + V - s;$$

and since from (5)
$$Q - \Sigma q_\lambda' = q + \nu - r + 1,$$
it follows that

$$q + \nu - r + 1 = g - 2n + 2 + V - s$$

or $\quad q = g - 2n + 1 - (\nu + s - r) + V,$

and finally by (3)

$$q = g - 2n + 1 - \tfrac{1}{2}(K - e) + V. \tag{8}$$

If there be no boundary-lines of the second kind, and hence if $V = 0$, this formula reduces to (6).

The circuits V of the boundary-lines of the second kind are, according to the preceding, to be counted in the pieces B which are removed, and in the way specified in § 55, namely: Each boundary-line is to be so described that the piece B lies on the left; and after B is extended in a plane, each circuit is to be counted as positive or negative, according as it is described in the direction of increasing or of decreasing angles.

We have yet to call attention to a special condition. It may happen that boundary-pieces, which belong either to different boundary-lines of the second kind or to one such line, meet in single points S. In such cases different conceptions are possible, both in regard to how a boundary-line shall be continued beyond a point S, and also in regard to the connection of the surface-pieces contiguous to S. Now formula (8) remains always valid, if we hold a conception once chosen. Yet, in order to remove all difficulties which may thereby occur, and in order to have something definite, we will assume that, when two boundary-pieces which belong to boundary-lines of the second kind meet in a point S, they are connected by an infinitely small cross-cut, *i.e.*, by an infinitely small boundary-line of the first kind. The advantage is thereby secured, that every boundary-line of the second kind without exception, if it be considered by itself, that is, apart from boundary-lines of the first kind which may possibly meet it, forms a *simply* closed line.

Formula (6) holds quite generally for surfaces which contain only boundary-lines of the first kind. Formula (8) on the other hand, for surfaces which possess both kinds of boundary-lines or only those of the second kind, holds only under the condition that the surface-pieces which are removed can be extended in planes. But if this condition be satisfied, then (8) remains equally valid, whether or not T itself can be extended in a plane. We will emphasize a case in which T can be so extended, and in which then formula (8) can be again reduced to the simple relation $q = g - U + 1$. If we assume that the complete surface W is closed at infinity, and if the case occur in which all the n points at infinity have been removed from T by means of boundary-lines of the second kind, which together make n circuits, then T can be extended in a plane.[1] If this case occur, the outer boundary-lines make n circuits, and the other $V - n$ circuits arise from the inner boundary-lines. The latter will, according to the hypothesis, be so described that the pieces B which are removed lie on the left, and T therefore on the right. But if we reverse the direction of description, in order to establish again the customary hypothesis that T lies on the left, each circuit at the same time changes its sign, and consequently

$$-(V-n) = n - V$$

[1] This is *perhaps* the only case in which T itself and also the pieces which were removed can be extended in planes; but it may be left undecided whether this cannot occur in still other cases.

is the number of positive circuits for these boundary-lines. But the case is different with the outer circuits. For a positive circuit (in the direction of increasing angles), in a piece which is removed and which contains a point at infinity, forms a negative circuit in T when that surface is extended in a plane. Therefore, if we also reverse here the circuit-direction, it remains a positive circuit. Thus the outer boundary-lines make n positive circuits, the inner boundary-lines $n - V$ such circuits, and consequently the boundary-lines of the second kind contribute
$$2n - V \qquad (9)$$
to U.

This value is increased by $+1$ by every cross-cut, for boundary-lines of the first kind (p. 229), while every slit leaves it unchanged; for a change of direction equal to $+\pi$ occurs at one end of a slit, and a change equal to $-\pi$ at the other. To determine the number of cross-cuts, which are contained only in boundary-lines of the first kind, we will divide these into two classes; let the first class include those which are connected with boundary-lines of the second kind, the second class all the others. The values of e and K which refer to these two classes may be denoted by e_1 and K_1, and e_2 and K_2 respectively; then
$$e_1 + e_2 = e, \quad K_1 + K_2 = K. \qquad (10)$$

Let us keep in mind, in reference to the first class, that when the sections discussed in § 50 are made in a line-system, the number of simple line-segments arising is always the same; namely,
$$\tfrac{1}{2}(e_1 + K_1),$$
even when the sections are so effected that simply closed lines arise. Hence we can so direct the sections in the line-system under discussion that all the boundary-lines of the second kind contained in it become simply closed lines. Then the boundary-lines of the first kind which are left form $\tfrac{1}{2}(e_1 + K_1)$ simple line-segments, and of these, since all the e_1 end-points lie in the interior, $\quad \tfrac{1}{2}(e_1 + K_1) - e_1 = \tfrac{1}{2}(K_1 - e_1)$
are non-dividing cross-cuts.

The second class of boundary-lines of the first kind, *i.e.*, those which are not connected with boundary-lines of the second kind, may form ρ distinct systems. If a boundary-point be assumed on each of the latter, they form by (2)

$$\tfrac{1}{2}(K_2 - e_2) + \rho$$

non-dividing cross-cuts. Since, however, every boundary-point represents a negative circuit, they furnish

$$-\rho + [\tfrac{1}{2}(K_2 - e_2) + \rho] = \tfrac{1}{2}(K_2 - e_2)$$

positive circuits. Consequently the number $2n - V$ found under (9) is to be increased by $\tfrac{1}{2}(K_1 - e_1)$ for the first class of boundary-lines of the first kind, and by $\tfrac{1}{2}(K_2 - e_2)$ for the second class. We thus obtain, with attention to (10), the value

$$U = 2n - V + \tfrac{1}{2}(K - e)$$

for the number U of positive circuits made by the aggregate of boundary-lines; by means of this relation (8) is reduced to

$$q = g - U + 1.$$

From the results of this paragraph we can now enunciate the proposition:

Every Riemann surface which possesses only a finite number of sheets and branch-points, the boundary-lines of which form a finite line-system (in the sense of § 50), can be modified into a simply connected surface by means of a finite number of cross-cuts.

57. We will now conclude these investigations by making another application, namely, *to the determination of the relation which exists between the number of corners, edges, and faces of an arbitrary body bounded by plane surfaces.*[1]

If we denote these numbers in order by e, k, and f, then, according to a proposition by Euler,

$$e - k + f = 2. \tag{1}$$

[1] F. Lippich, "Zur Theorie der Polyeder," *Sitz.-Ber. d. Wien. Akad.*, Bd. 84, Abth. II., Juni-Heft, 1881.

SIMPLY AND MULTIPLY CONNECTED SURFACES. 243

But this relation does not hold for every arbitrarily formed body with plane faces; on the contrary, a number which depends upon the order of connection both of the aggregate of surfaces, and also of the individual lateral faces, must in general be added to the right side. For instance, the Eulerian relation does not hold for the body represented in Fig. 52, in which a smaller parallelopiped so rests upon a larger that the face of the smaller covers a portion of the interior of a face of the larger. We can at once convince ourselves of this by an enumeration. For in this case $e = 16$, $k = 24$, $f = 11$; therefore

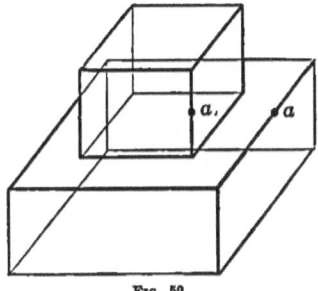

Fig. 52.

$$e - k + f = 3, \text{ and not } 2,$$

as the Eulerian relation requires. In like manner this relation does not always hold if there be a cavity in the body, or if it be closed after the manner of a ring.

We will now assume that the aggregate of surfaces of the body is $(q + 1)$-ply connected; that therefore q cross-cuts modify it into a simply connected surface. Since this surface is closed, we must, by § 46, assume a boundary-point. Let this be denoted by a, and be situated on an edge (Fig. 52). Now the edges form a line-system in the surfaces of the body. This can either be wholly connected or consist of distinct parts. Let the number of such parts be n, where n can also be equal to unity. This line-system could, by § 51, have been regarded as a system of cross-cuts, if it had satisfied conditions (1) and (2), given in that paragraph. Condition (1) is indeed satisfied, since the lines possess no end-points in the interior of the surface; but not (2), since, in case n be not equal to unity, the lines are not all connected with the boundary-point a. Nevertheless we can cause this condition to be satisfied, by also assuming one boundary-point on an edge belonging to each of the other $n-1$ parts of the system of edges, just as the boundary-point a was assumed in one of those parts. Let these points be denoted by $a_1, a_2, \cdots, a_{n-1}$. (In Fig. 52 it is necessary to assume only one such point, a_1.) Then the surface possesses n boundary-points, and every line is connected with some one boundary-point; consequently condition (2) is satisfied. Therefore, by § 53, the line-system which consists of the edges, now forms a system of cross-cuts, quite definite in number; let this number be s.

But now, after $n-1$ new boundary-points are taken out of the surface, its order of connection is increased by $n-1$. Thus $q + n - 1$ cross-cuts are necessary to change it into a simply connected surface. If we imagine the surface to be cut through along the edges, which form s cross-cuts, we resolve it into distinct pieces; namely, into the individual bounding-faces of the body, the number of which was f. These are not, in general, all simply connected. (In Fig. 52 one was not, namely, that one upon which the smaller body rests.) If we denote by p the total number of cross-cuts which are necessary to make all the bounding-faces simply connected, and if we add these cross-cuts, none of which divides a face, we again obtain f distinct pieces; these pieces are now, however, all simply connected. Consequently we have: The $(q+n)$-ply connected aggregate of surfaces of the body, after the removal of the $n-1$ boundary-points, is resolved by $s+p$ cross-cuts into f distinct pieces, each of which is by itself simply connected.

But now, on the other hand, we can first change the same surfaces into *one* simply connected surface by means of $q + n - 1$ cross-cuts, and then resolve this surface into f distinct pieces by means of $f-1$ additional cross-cuts. The former surface is therefore also resolvable into f distinct pieces, each by itself simply connected, by means of

$$(q + n - 1) + (f - 1)$$

cross-cuts. Consequently, according to Riemann's fundamental proposition,
$$(q+n-1)+(f-1) = s+p,$$
or
$$n-s+f = 2+p-q. \qquad (2)$$

The number $n-s$ which appears in this formula can be expressed in terms of the numbers e and k of corners and edges. For at every corner at least three edges meet, and hence each corner forms a nodal-point of the line-system which consists of these edges; and if we denote by e_3, e_4, e_5, ... the numbers of corners in which 3, 4, 5, ... edges meet, we have
$$e = e_3 + e_4 + e_5 + \cdots.$$

If, further, we count all the edges which meet in the individual corners, we obtain double the number of all the edges, since every edge is counted twice. Hence
$$2k = 3e_3 + 4e_4 + 5e_5 + \cdots.$$

If we now wish to resolve the system of edges into the s cross-cuts of which it consists, we must make the sections discussed in § 50; then the s cross-cuts appear as s simple line-segments, the number of which is half as great as the number of their end-points. Therein, by § 51, each of the n points a, a_1, a_2, ..., a_{n-1}, must be regarded as forming two end-points. Since, in addition, each corner in which h edges meet, as an h-ple nodal-point, furnishes $h-2$ end-points, we obtain
$$2s = 2n + e_3 + 2e_4 + 3e_5 + \cdots.$$

If the preceding expression for $2k$ be subtracted from this, we get
$$2s - 2k = 2n - 2(e_3 + e_4 + e_5 + \cdots)$$
$$= 2n - 2e;$$
consequently
$$n - s = e - k,$$
and from (2)
$$e - k + f = 2 + p - q. \qquad (3)$$

This is the desired relation. In the general case, therefore, the number $p - q$ is added to the number 2 on the right side of the Eulerian relation (1); in this q is the number of cross-cuts necessary to modify the aggregate of surfaces into a simply connected surface, and p the number necessary to that end in all the individual boundary-faces.

The Eulerian relation therefore holds only when $p = q$. In an ordinary polyedron, everywhere convex, this is in fact the case, because then $p = q = 0$. For some special cases, and the way in which the numbers e, k, f must be counted in order that equation (3) may remain valid, we refer to the dissertation cited above.

SECTION X.

MODULI OF PERIODICITY.[1]

58. Let $f(z)$ denote an arbitrary *algebraic* function. Let us conceive as the region of the variable z a surface consisting of as many sheets and containing such branch-points as the nature of this function $f(z)$ requires. We will surround with small closed lines the points of discontinuity of this function and thus exclude them. We will assume provisionally that all the points of discontinuity are enclosed in this way, but we shall very soon see that certain kinds of points of discontinuity need not be excluded. We will call the surface so formed T. This now possesses a finite order of connection, and can therefore, if it be multiply connected, be modified into a simply connected surface by means of a finite number of cross-cuts. For, since the function in question is an algebraic one, this is, by § 54, at all events the case before the exclusion of the points of discontinuity. But since an algebraic function possesses only a finite number of points of discontinuity (§ 38),

[1] The special investigation of the logarithmic and exponential functions given in § 22 and § 23 may serve as illustrations of the general considerations contained in this section. Other examples will be found in § 61.

MODULI OF PERIODICITY. 247

therefore by the exclusion of these points only a finite number of new boundary-lines are added; accordingly by § 56 the order of connection also remains finite after the exclusion of the points of discontinuity. Consequently, if the surface T be multiply connected, we will modify it into a simply connected surface by means of cross-cuts, and designate the new surface by T'. Then every closed line in T' forms the complete boundary of a portion of the surface, in which $f(z)$ is finite and continuous. Hence, if the function defined by the integral

$$w = \int_{z_0}^{z} f(z)dz$$

be formed by integrating from an arbitrary fixed initial point z_0 to a point z, along an arbitrary path which lies wholly within T', then any two such paths together make a closed line, and this line bounds completely a portion of the surface in which $f(z)$ is everywhere continuous; therefore w acquires at z, along all such paths, one and the same value (§ 18). Consequently w is a function of the upper limit z, and remains uniform everywhere within T'.[1]

[1] The case in which two paths taken together form a closed line which intersects itself is no exception to the above. For we can always resolve such a line into several simply closed lines. (Cf. Fig. 53.) The resolution is effected in the following way: Whenever, in tracing the line from an arbitrary point z_0, we have returned to a point already once passed (e.g., a), and thus have traced a simply closed line (e.g., $abcda$), we separate this and regard the part which follows (e.g., ae) as the continuation of the part (z_0a) which preceded the part separated. If this mode of procedure be repeated as often as the same condition arises, there is finally left a line likewise simply closed; and in this way the given line is resolved into

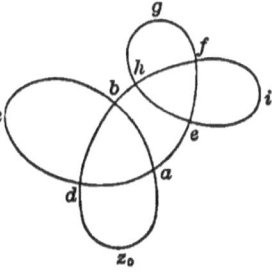

FIG. 53.

several simply closed lines. (In the figure the lines which are separated are $abcda$ and $efghe$, and that which is left is $z_0aeifhbdz_0$.) The above

But the case is different when we consider the function w in the surface T, and when therefore we let the path of integration cross the cross-cuts. In order to examine this, we will first direct our attention to the case in which no cross-cut

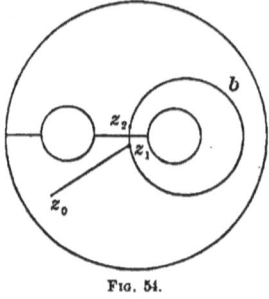

Fig. 54.

is divided into segments by a subsequent one which starts from it. Now both edges of each cross-cut belong to the boundary of T', so that these are connected, and we can draw a closed line b, running entirely in the interior of T', which leads from one edge of the cross-cut to the other edge of the same. Let z_1 and z_2 (Fig. 54) be two points lying infinitely near each other on opposite sides of the cross-cut. We will now inquire whether

$$w = \int_{z_0}^{z_1} f(z) dz,$$

when the paths of integration still run entirely in T', acquires at z_1 and z_2 values that are equal (accurately speaking, different by an infinitesimal quantity) or different. But if we denote the values of w at z_1 and z_2 by w_1 and w_2 respectively, we have

$$w_2 = \int_{z_0}^{z_2} f(z) dz = \int_{z_0}^{z_1} f(z) dz + \int_{z_1}^{z_2} f(z) dz,$$

the first integral to be taken along an arbitrary path running in T', the second along a closed line b leading from z_1 to z_2 within T'. Thus

$$w_2 - w_1 = \int_{z_1}^{z_2} f(z) dz.$$

integral, extended along the simply closed lines, is now equal to zero, and therefore it is also zero taken along the given line, since this integral is equal to the sum of the preceding. Then, if the given path be formed by two paths leading from z_0 to z, the integral has the same value along both paths (§ 18).

Hence w_1 and w_2 have the same or different values according as the integral

$$\int f(z)\, dz,$$

extended along the closed line b, is zero, or has a value A different from zero. In the first case w remains continuous on crossing the cross-cut; in the latter case w springs abruptly from w_1 to $w_2 = w_1 + A$, and is therefore discontinuous. But this abrupt change is the same at all places of the same cross-cut, because the value of the integral does not change, if we enlarge or contract the closed line b in such a way that it begins and ends at two other infinitely near points on opposite sides of the same cross-cut (§ 19). This quantity A, which is thus *constant* along the entire cross-cut, and by which the function-values on one side of the cross-cut exceed those on the other, is called the *modulus of periodicity* corresponding to this cross-cut. The case is exactly similar for every cross-cut, because the two edges of each one are connected, and therefore a closed line can be drawn from a point on one side to an infinitely near point on the other side through the interior of T'. Thus to every cross-cut corresponds a modulus of periodicity, which remains constant for one and the same cross-cut (yet always under the hypothesis that no cross-cut is divided into segments by a subsequent one). But if we now assume that the function w proceeds *continuously* in T also, and hence also over the cross-cut, it acquires at z_1, on the path $z_0 z_1 b z_2 z_1$, which crosses the cross-cut, a value greater by the modulus of periodicity than the value acquired on the path $z_0 z_1$, which does not cross the cross-cut. For in the former case the value of w at z_1 is regarded as the uninterrupted continuation of w_2, while on the second path w acquires the value w_1, and

$$w_2 = w_1 + A.$$

There occurs here a condition similar to that which we found to exist in the case of branch-cuts (cf. § 13), and as long as the surface T consists of only a single sheet, we can also regard every cross-cut as actually a branch-cut, over which the surface continues into another sheet. But we must then suppose that infinitely many sheets lie one below another, since, for every new passage of the cross-cut, the value of the function w is increased by A, and the original value never occurs again. If the surface T itself already consist of several sheets, that mode of representation would indeed be possible, but yet it would be too complicated, and hence would offer no real advantage.

The sign of A changes if the closed line b be described in the opposite direction; but we will always so assume the modulus of periodicity that it is equal to the integral taken along the closed line b in the direction of increasing angles.

If we now conceive all possible paths which lead from an initial point z_0 to an arbitrary point z through the interior of T, then these paths can either cross none of the cross-cuts or intersect one or more cross-cuts one or more times. Hence w can acquire at one and the same point z very different values, according to the nature of this path, and it is therefore a

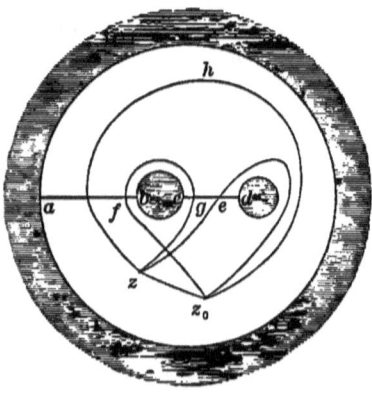

FIG. 55.

multiform function of the upper limit of the integral. But since this diversity of values of w at the point z is due solely to the passages over the cross-cuts, these different values can differ from one another only by multiples of the moduli of periodicity. Hence, if $A_1, A_2, A_3, \cdots$ denote the moduli of periodicity for the single cross-cuts, $n_1, n_2, n_3, \cdots$ positive or negative integers, and w and w' two different values of w at the point z, then

$$w' = w + n_1 A_1 + n_2 A_2 + n_3 A_3 + \cdots.$$

An example may make this clear. Fig. 55 represents a triply connected surface; let the cross-cuts be ab and cd, and let the moduli of periodicity for the same be A_1 and A_2, respectively, so taken that the passage from one side of the cross-cut to the other side along a closed line is made in the direction of increasing angles. If we designate the value acquired by the function w on a path by adding the path in brackets to the letter w, we have

$$w(z_0 e z) = w(z_0 z) + A_2,$$
$$w(z_0 f g z) = w(z_0 z) - A_1 + A_2,$$
$$w(z_0 h z) = w(z_0 z) + A_1.$$

From this it is evident that the function defined by the integral

$$w = \int_{z_0}^{z} f(z)\, dz$$

possesses a multiformity of a quite peculiar kind; namely, that the different values which it can acquire for the same value of z differ from one another only by multiples of constant quantities. If we now take the inverse function, *i.e.*, if we regard z as a function of w, then this is a periodic function, since it remains unchanged when we increase or diminish the argument w by arbitrary multiples of the moduli of periodicity. By this also the name *modulus of periodicity* is justified, since we can say, analogously to the language of the theory of numbers, that z acquires equal values for such values of w as are congruent with one another to a modulus of periodicity, i.e., as have a difference equal to a multiple of the modulus of periodicity.

59. We have hitherto assumed that the cross-cuts are so drawn that no one of them is divided into segments by a subsequent cross-cut which starts from it. But if one be so divided, as for instance in Fig. 56, where the one cross-cut ad is divided by the second ce into the two segments ac and cd, the modulus of periodicity B_1 of the one segment ac may possibly differ from that B_2 of the other segment cd. For B_1 is equal to the integral $\int f(z)dz$ taken along the line b_1, B_2 is equal to the same integral taken along b_2. If these integrals have different values, then the moduli of periodicity B_1 and B_2 are different. Thus

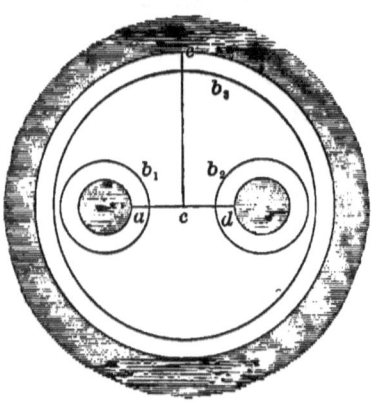

Fig. 56.

the modulus of periodicity does not now remain constant along an entire cross-cut, but only from one node of the net of cuts to the next. But now a modulus of periodicity B_3 corresponds to the cross-cut ce, and hence there are three moduli of periodicity, notwithstanding that only two cross-cuts are necessary to modify our surface into a simply connected surface. But in such a case there always exist relations between the single moduli of periodicity. In our example the integral taken along b_3 is equal to the sum of the integrals taken along b_1 and b_2 (§ 19), and hence

$$B_3 = B_1 + B_2;$$

thus we have in fact only two moduli of periodicity which are

MODULI OF PERIODICITY. 253

independent of each other, *i.e.*, just as many as there are cross-cuts.

To prove now in general that there are always only as many moduli of periodicity independent of one another as there are cross-cuts, we observe that the cross-cuts in most cases can be drawn in various ways. But there is always one mode of resolution in which no cross-cut is divided into segments by a subsequent cross-cut. This is always effected by beginning every cross-cut at a point of the original boundary and also ending it at such a point. If the surface be closed and hence possess only a single boundary-point (§ 46), we have only to begin and end each cross-cut at this point.

Now let an $(n+1)$-ply connected surface first be so resolved into a simply connected surface by means of n cross-cuts that thereby no cross-cut is divided into segments by another; we then have, for this mode of resolution, exactly as many moduli of periodicity as cross-cuts. Let these be

$$A_1, A_2, \cdots, A_n.$$

Next let the same surface be resolved in another arbitrary way. Thereby the single cross-cuts are divided into segments with different moduli of periodicity, and the number of the latter is greater than n; let these be

$$B_1, B_2, \cdots, B_m \ (m > n).$$

Now let the variable z describe from any arbitrary point z_0 a closed line which crosses only one cross-cut of the first system, and let the modulus of periodicity for this cross-cut be A_k; then, if w_0 and w denote the values of the function at the beginning and after the completion of the closed line, we have

$$w = w_0 + A_k.$$

But if we now suppose the surface to be resolved in the second way, the same closed line may cross several cross-cuts of the second system; hence by § 58 the value of w must be obtained also in the form

$$w = w_0 + h_1 B_1 + h_2 B_2 + \cdots + h_m B_m,$$

wherein h denotes a positive or negative integer (zero included). Consequently

$$A_\lambda = h_1 B_1 + h_2 B_2 + \cdots + h_m B_m.$$

Now, conversely, let the variable z describe from z_0 a closed line which crosses only one cross-cut of the second system, and let the modulus of periodicity of this cross-cut be B_λ; then the final value of the function is first

$$w_0 + B_\lambda;$$

but, if the crossings of the cross-cuts of the first system be considered, that value is also obtained in the form

$$w_0 + g_1 A_1 + g_2 A_2 + \cdots + g_n A_n,$$

wherein g likewise denotes a positive or negative integer (zero included). From this follows

$$B_\lambda = g_1 A_1 + g_2 A_2 + \cdots + g_n A_n.$$

Consequently we obtain between the two systems of the moduli of periodicity A and B the following two sets of equations:

$$(1) \begin{cases} A_1 = h_1' B_1 + h_2' B_2 + \cdots + h_m' B_m \\ A_2 = h_1'' B_1 + h_2'' B_2 + \cdots + h_m'' B_m \\ \cdots \cdots \cdots \cdots \cdots \cdots \cdots \cdots \\ A_n = h_1^{(n)} B_1 + h_2^{(n)} B_2 + \cdots + h_m^{(n)} B_m \end{cases}$$

and

$$(2) \begin{cases} B_1 = g_1' A_1 + g_2' A_2 + \cdots + g_n' A_n \\ B_2 = g_1'' A_1 + g_2'' A_2 + \cdots + g_n'' A_n \\ \cdots \cdots \cdots \cdots \cdots \cdots \cdots \cdots \\ B_m = g_1^{(m)} A_1 + g_2^{(m)} A_2 + \cdots + g_n^{(m)} A_n. \end{cases}$$

Since now according to the assumption $m > n$, we can eliminate the n quantities A from equations (2) and thereby obtain $m - n$ relations between the quantities B. But since we can also obtain these relations by substituting in (2) the values of A from (1), they must be homogeneous linear equations with integral coefficients. Therefore we conclude: If previous cross-cuts be divided by subsequent cross-cuts into segments which have different moduli of periodicity, so that

MODULI OF PERIODICITY.   255

in all m moduli of periodicity exist, while only n cross-cuts occur, then there are $m - n$ linear homogeneous equations of condition with integral coefficients between these m moduli of periodicity, and of these moduli only n, *i.e.*, just as many as there are cross-cuts, are independent of one another.

We can also, without any calculation, reach the same conclusion by a simple consideration. For, after the surface has been made simply connected by means of cross-cuts, its boundary can be traced in a continuous description (§ 53, IX.). The cross-cuts and their segments enter in this description in a definite succession. If, for each cross-cut, the modulus of periodicity be known for that segment at which we arrive first in the description, then the moduli of periodicity for the other segments are given by linear relations. We will show this only in an example.

In the quadruply connected surface represented by Fig. 57, let ab, cd, ef be the three cross-cuts which modify the surface into a simply connected surface. Let the letters p, q, r, s, t, u, v, x, y, z denote the values acquired by the function w at the corresponding points which are situated infinitely near the cross-cuts. If we now describe the cross-cuts, together with the original boundary, in the direction $aefc \cdots$, let the moduli of periodicity be known for the three segments ae, ef, fc, and be denoted by

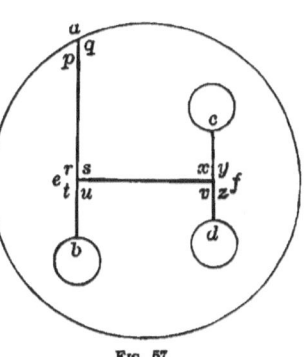

Fig. 57.

$$q - p = s - r = A_1, \quad s - u = x - v = A_2, \quad x - y = A_3;$$

we then wish to obtain the moduli of periodicity for the segments eb and fd, and will denote these by

$$u - t = X_1, \quad z - v = X_2.$$

To find these, we remark that continuity exists between the function-values at any two consecutive points which are not separated by a cross-cut; that their difference is therefore infinitesimal. Consequently we can let

Thus we obtain $\quad t - r = 0, \; z - y = 0.$

$$X_1 = u - t = u - r = (s - r) - (s - u) = A_1 - A_2,$$
$$X_2 = z - v = y - v = (x - v) - (x - y) = A_2 - A_3,$$

by which X_1 and X_2 are expressed in terms of A_1, A_2, A_3.

60. We have hitherto assumed that all the points of discontinuity are removed from the z-surface by means of small enclosures, so that the function $f(z)$ remains finite in the surface T so formed. But we will now show that it is in fact not necessary to exclude all the points of discontinuity, and will inquire for what points the enclosures need not be drawn.

The modulus of periodicity A for a particular cross-cut is, as was shown in § 58, the value of the integral $\int f(z)dz$, extended over a closed line b which leads from one side of the cross-cut through the interior of the simply connected surface T' to the other side of the same cross-cut. But this integral in many cases may have the value zero. Let us assume that the closed line b encloses a place removed from the z-surface which contains a point of discontinuity a (which is not at the same time a branch-point) of the function $f(z)$. Then by § 42 the integral $\int f(z)dz$ has a value different from zero only when the term

$$\frac{c'}{z - a}$$

is present in the expression which indicates how $f(z)$ becomes infinite; in all other cases the integral has the value zero. For instance, the integral equals zero when $f(z)$ is infinite at a as

$$\frac{c''}{(z - a)^{z'}}$$

or as

$$\frac{c^{(n)}}{(z - a)^n} + \frac{c^{(n+1)}}{(z - a)^{n+1}} + \cdots$$

is infinite, wherein n denotes a positive integer different from unity. In such a case the function w remains continuous on crossing a cross-cut; hence it is not necessary to exclude the point of discontinuity, and the cross-cut need not be considered. If we assume, for instance, a simply connected piece of the z-surface, in which are contained only points of discontinuity of the kind in question, then the integral $\int f(z)dz$ acquires the same value along two paths which enclose such a point of discontinuity, because this integral, taken round the point of discontinuity, has the value zero (§ 18). Hence, in such a piece of the surface, the function

$$w = \int_{z_0}^{z} f(z)dz$$

is likewise a uniform function of the upper limit, just as if the piece of the surface contained no point of discontinuity at all.

This is one kind of point of discontinuity which need not be excluded. Let us now turn to branch-points. The integral $\int f(z)dz$, taken along the closed line b, has the value zero when this line encloses a winding-point of the $(m-1)$th order at which $f(z)$ becomes infinite of an order not higher than $\frac{m-1}{m}$ (§ 21); and, in general, when the term which is infinite of the first order is wanting in the expression which indicates how $f(z)$ becomes infinite at the branch-point (§ 42). In this case, therefore, the discontinuity- and branch-point need not be excluded, and thus it is likewise unnecessary to consider the cross-cut. But we remark that, since the z-surface now consists of several sheets, it may be multiply connected without the exclusion of points of discontinuity. Thus cross-cuts will always in such cases be required in order to modify the surface into a simply connected surface, and to these will correspond moduli of periodicity.

Finally, we can also determine in what case the point at infinity must be excluded. The value of the integral, for a

line enclosing the point $z = \infty$, depends upon the nature of the function
$$z^2 f(z)$$
for $z = \infty$ (§ 43). Thus this point must be excluded when
$$\lim [zf(z)]_{z=\infty} \text{ is finite, and not zero;}$$
and in general when, and only when, in the development of $f(z)$ in ascending and descending powers of z, a term of the form
$$\frac{g}{z}$$
is present.

If now, for a given function $f(z)$, all those points have been excluded from the z-surface which must necessarily be excluded, and only these, then, *within the surface T so formed, the integral $\int f(z)dz$, taken along a closed line which forms by itself alone the complete boundary of a portion of the surface, is always equal to zero.*

For the portion of the surface so bounded contains then either no points of discontinuity at all, or only such as lead to the value zero for the integral taken along the boundary. In this it is, of course, assumed that the closed line does not pass through a point of discontinuity or a branch-point.

61. We will now apply the preceding considerations to some examples.

1. *The Logarithm.*

We will recall first the function $\log z$, or the function defined by the integral
$$w = \int_1^z \frac{dz}{z},$$
already discussed in § 22 and § 23. In this $f(z) = \frac{1}{z}$ is uniform, and hence the z-surface consists of one sheet. Further, $z = 0$ is a point of discontinuity, and
$$\lim [zf(z)]_{z=0} = \lim \left(z \cdot \frac{1}{z}\right)_{z=0} = 1.$$

Hence this point must be excluded. If we now assume that the z-surface is closed at infinity, the point $z = \infty$ must also be excluded, because
$$\lim \left[zf(z) \right]_{z=\infty} = 1.$$
By the exclusion of these two points, the surface T is made doubly connected, and a cross-cut which connects the circles enclosing the two points 0 and ∞ modifies it into a simply connected surface (Fig. 58).
The modulus of periodicity A is equal to the value of the integral
$$\int \frac{dz}{z},$$
taken along a closed line, which makes a circuit round the origin in the direction of increasing angles, and hence

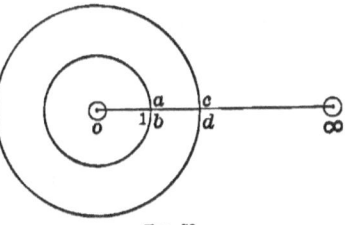

Fig. 58.

$$A = 2\pi i.$$

Such a line also encloses the point ∞ at the same time, and for this we obtain (§ 43)
$$\int \frac{dz}{z} = -2\pi i \lim \left[\frac{z^2 \frac{1}{z}}{z} \right]_{z=\infty} = -2\pi i,$$
if the integration be extended in the positive boundary direction, and if, therefore, the cross-cut be crossed in a direction opposite to the former.

2. *The Inverse Tangent.*

$$w = \int_0^z \frac{dz}{1+z^2}.$$

Here $\quad f(z) = \dfrac{1}{1+z^2}$

is likewise uniform and becomes infinite of the first order, for $z = i$ and $z = -i$; on the other hand,

$$\lim \left[zf(z)\right]_{z=\infty} = \lim \left[\frac{z}{1+z^2}\right]_{z=\infty} = 0.$$

Hence we need exclude only the points $z = i$ and $z = -i$ by means of small circles (Fig. 59), and we then obtain, assuming the z-surface to be closed at infinity, a doubly connected surface; this is changed into a simply connected surface by a cross-cut which joins the small circles round $+i$ and $-i$. The modulus of periodicity A is the value of the integral

$$\int dw,$$

taken along a closed line which makes a circuit round the point $+i$ in the direction of increasing angles, and hence, as we have already found in § 20,

FIG. 59.

$$A = \pi.$$

The same line can be regarded as one which makes a circuit round the point $-i$ in the direction of the decreasing angles, and it then furnishes the same modulus of periodicity.

If we now assume that the z-surface is not closed at infinity, but is bounded by a closed line which we then enlarge indefinitely, the surface T becomes triply connected when the two points $+i$ and $-i$ are excluded. Therefore, two cross-cuts are in this case necessary to change the surface into one simply connected. But now, since the integral

$$\int \frac{dz}{1+z^2},$$

taken along a closed line, has the value $+\pi$ or $-\pi$ or 0, according as the line makes a circuit round $+i$ or $-i$ or both, in the direction of increasing angles (§ 20), the moduli of periodicity in reference to the two cross-cuts have the values $+\pi$ and $-\pi$, or the one has the value $\pm \pi$ and the other the value zero, according to the mode of drawing the cross-cuts.

Hence the function $w = \text{arc-tan } z$ also changes here by multiples of π.

The inverse function $z = \tan w$ is now periodic with the period π. The representation of the z-surface, assumed to be closed at infinity, on the w-surface, is here made in a way exactly similar to that shown in § 23 for the exponential function; in place of the circles enclosing the points 0 and ∞ there enter here only those which enclose the points $+i$ and $-i$. If we assume that the cross-cut which joins these circles runs along the ordinate axis, the w-surface is divided into strips bounded by straight lines which run parallel to the ordinate axis, and which pass through the points $0, \pm \pi, \pm 2\pi, \pm 3\pi, \cdots$ (Fig. 60). In each of these strips the function $z = \tan w$ acquires all its values, and, indeed, each but once, because, except as to multiples of the modulus of periodicity, only one value of w corresponds to each value of z, the z-surface consisting of only one sheet.

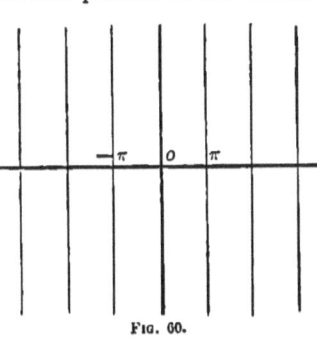

FIG. 60.

We will now examine this function in the inverse manner, by commencing with the periodic function. If $z = \phi(w)$ denote a uniform simply periodic function with the modulus of periodicity A, that is, a uniform function which possesses, the property that

$$\phi(w + A) = \phi(w),$$

then the w-surface can be so divided into strips that the function acquires all its values in each strip, and has the same value at every two points situated in different strips which differ by A or a multiple of A (Fig. 61). For, if we draw any line BC which does not intersect itself, the points $w + A$, which are obtained from the points of the line BC by adding A, form a line DE parallel to the line BC. Thus the function ϕ

has the same values along DE as along BC. The same is true of all lines which run parallel to these at equal distances.

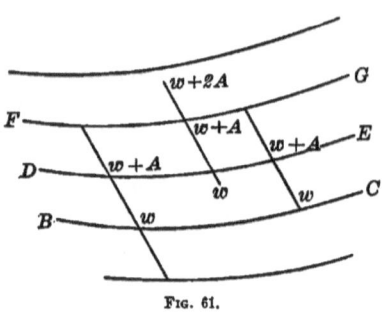

Fig. 61.

Moreover, if w be a point in the interior of the strip $BCDE$, then $w + A$ lies in the interior of the adjacent strip $DEFG$, $w + 2A$ in the interior of the next following strip, etc. Hence at these points the function again has the same value. Now, since every two points, w and $w + nA$, at which the function has the same value, lie in different strips, it must acquire all its values in each strip.

We will now assume further that the function $z = \phi(w)$ becomes infinite of the first order at only *one* finite point $w = r$ in one and the same strip; we can then show that it also becomes zero only once in every strip and hence acquires each value only once. To this end let the points at which $\phi(w)$ becomes zero within the strip considered be denoted by $s, s', s'', \ldots$, and let the number of these points be n and assume that none of them lies at infinity. If we now draw, from two points w and $w + c$ situated on one of the two lines which bound the strip, straight lines to the points $w + A$ and $w + c + A$, situated on the other bounding-line (Fig. 62), we obtain a parallelogram with vertices w, $w+c$, $w + c + A$, $w + A$; and if, as was assumed, the points r, s, $s', s'', \ldots$ all lie in the finite part of the surface, we can always so choose the points w and $w+c$ that $r, s, s', s'', \ldots$ lie within the parallelogram. If we now

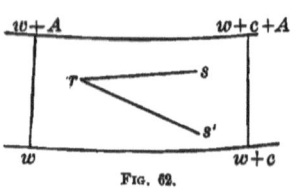

Fig. 62.

take the integral $\int d \log \phi(w)$ along the boundary of this parallelogram, we obtain by § 35, (1),

$$\int d \log \phi(w) = 2\pi i(n - 1),$$

since $\phi(w)$ becomes n times zero and once infinite within the parallelogram. This integral may be divided into four parts, taken along the four sides of the parallelogram. But we remark that $\int d \log \phi(w)$ is independent of the path of integration as long as this does not cross one of the lines rs, rs', etc., each of which connects points at which $\phi(w)$ becomes infinite or zero (§ 22). If we take it along the straight line which leads from $w + A$ to w, it acquires the value zero; for in the first place it is equal to $\log \phi(w) - \log \phi(w + A)$, and since none of the lines rs is crossed, not only is $\phi(w + A) = \phi(w)$, but also $\log \phi(w + A) = \log \phi(w)$. [If one line rs were crossed, we should have $\log \phi(w + A) = \log \phi(w) \pm 2\pi i$.] For the same reason the integral which is taken along the straight line leading from $w + c$ to $w + c + A$ is also zero. But along the two lines which bound the strip from w to $w + c$, and from $w + A$ to $w + c + A$, $\log \phi(w)$ passes through the same values, and since these lines are described in opposite directions, the integrals taken along them cancel each other. Consequently the integral in the preceding equation, to be taken along the entire boundary of the parallelogram, is equal to zero, and therefore

$$n = 1.$$

Hence the function $\phi(w)$ becomes zero only once in the strip considered. But then it can also acquire any arbitrary value k only once in the same strip; for, if we form the function $\phi(w) - k$, this is periodic just as $\phi(w)$ is periodic, and it becomes infinite only once for $w = r$ just as $\phi(w)$ does; therefore it also becomes zero only once in the same strip, i.e., $\phi(w)$ becomes equal to k only once.

We can now, by § 29, let

(1) $$z = \phi(w) = \frac{c}{w - r} + \psi(w),$$

wherein c denotes a given constant, and $\psi(w)$ a function which no longer becomes infinite in the strip to be considered, but only in the other strips. From this follows

(2) $$\frac{dz}{dw} = \phi'(w) = -\frac{c}{(w-r)^2} + \psi'(w).$$

Since now $\psi'(w)$ remains finite everywhere in the strip, therefore $\frac{dz}{dw}$ also becomes infinite only for $w = r$, i.e., only where z becomes infinite, and this result must hold in like manner for all the strips. But while z is infinite of the first order, $\frac{dz}{dw}$ is infinite of the second order. Hence, if we regard $\frac{dz}{dw}$ as a function of z, it is infinite only for $z = \infty$, and then of the second order. Since, moreover, z acquires each value only once in one and the same strip, there corresponds only one value of w to each value of z, in one and the same strip. Consequently w is a function of z which has indeed an infinite number of values for each value of z, but these values differ from one another only by multiples of the modulus of periodicity, i.e., by constant quantities. Accordingly $\frac{dw}{dz}$ is a uniform function of z, since the constants vanish in the differentiation. Hence the reciprocal function $\frac{dz}{dw}$ must likewise be a uniform function of z. If we combine this with the preceding results, it follows that $\frac{dz}{dw}$ is a uniform function of z, which becomes infinite only for $z = \infty$, and here of the second order. Consequently $\frac{dz}{dw}$ is an integral function of z of the second degree (§ 31). Such a function must by § 36 also twice acquire the value zero. If we denote by a and b the values of z for which this occurs, and by C a constant, we have

(3) $$\frac{dz}{dw} = C(z-a)(z-b),$$

and hence

$$w = \int \frac{dz}{C(z-a)(z-b)}.$$

Therefore a simply periodic function, which becomes infinite of the first order only for one finite point in each strip, is the inverse function of the preceding algebraic integral.

The quantities a and b cannot have equal values in this integral, for in that case the function

$$w = \int \frac{dz}{C(z-a)^2}$$

would be a uniform function of the upper limit (§ 60), and then z could not be a periodic function.

The constant C can be expressed in terms of c; for from (3) we get

$$C = \lim \left[\frac{\frac{dz}{dw}}{z^2} \right]_{z=\infty},$$

and with help of equations (1) and (2)

$$C = \lim \left[\frac{\frac{-c}{(w-r)^2} + \psi'(w)}{\left\{ \frac{c}{w-r} + \psi(w) \right\}^2} \right]_{w=r},$$

or $\quad C = \lim \left[\dfrac{-c + (w-r)^2 \psi'(w)}{\{c + (w-r)\psi(w)\}^2} \right]_{w=r} = -\dfrac{1}{c}.$

We then have

$$w = \int \frac{-c\,dz}{(z-a)(z-b)}.$$

The modulus of periodicity A is equal to the value of this integral, taken along a closed line which encloses either the point a or the point b. If we integrate round a in the direction of increasing angles, we obtain

$$A = 2\pi i \lim \left[\frac{-c(z-a)}{(z-a)(z-b)} \right]_{z=a} = \frac{2\pi i c}{b-a};$$

for integration round b we should obtain the opposite value. If we assign the value h to the lower limit, i.e., if z acquire the

266   THEORY OF FUNCTIONS.

value h at the point $w = 0$, we have, since for $w = r$ and $w = s$, $z = \infty$ and $z = 0$ respectively,

$$r = \int_h^\infty \frac{-c\,dz}{(z-a)(z-b)}, \qquad s = \int_0^{'h} \frac{+c\,dz}{(z-a)(z-b)}.$$

3. *The Inverse Sine.*

$$w = \int_0^z \frac{dz}{\sqrt{1-z^2}}.$$

Here the z-surface for the function

$$f(z) = \frac{1}{\sqrt{1-z^2}}$$

consists of two sheets. We have the two branch-points $z = +1$ and $z = -1$, which are at the same time points of discontinuity. But these points need not be excluded, since $f(z)$ becomes infinite at them only of the order $\frac{1}{2}$; on the other hand the point $z = \infty$ must be excluded, because

$$\lim \left(\frac{z}{\sqrt{1-z^2}} \right)_{z=\infty} = \frac{1}{\sqrt{-1}}$$

is finite, and in fact the point ∞ must be excluded in both sheets, since it is not a branch-point. For this reason the connection of the surface in this example remains the same, whether we assume that the two sheets of the z-surface are closed at infinity, or imagine a closed line drawn in each sheet as a boundary, and then enlarge these lines indefinitely. In Fig. 63 the latter mode of representation is chosen on account of its greater practicability. The branch-cut is drawn from -1 to $+1$,

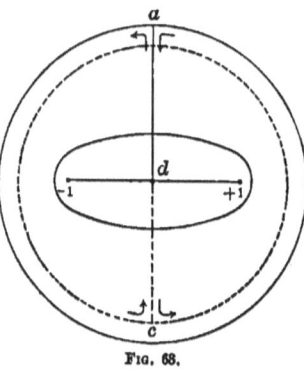

Fig. 63.

and the lines running in the second sheet are dotted. This surface, T, is doubly connected, and the cross-cut, in order not to divide the surface, must cross the branch-cut. It is denoted by the line adc, the part dc of which runs in the second sheet. The modulus of periodicity is the value of the integral

$$\int \frac{dz}{\sqrt{1-z^2}},$$

taken in the direction of increasing angles along a closed line which encloses the two points -1 and $+1$; this line may be drawn either in the first or in the second sheet. If we assume that the positive sign is to be attached to the radical at the points which lie in the first sheet in the immediate vicinity of the branch-cut, and on the left side of the same taken in the direction from -1 to $+1$, and if we let the closed line run in the first sheet, we can contract this line up to the branch-cut, and we then have

$$A = \int_{+1}^{-1} \frac{dz}{\sqrt{1-z^2}} - \int_{-1}^{+1} \frac{dz}{\sqrt{1-z^2}} = -2 \int_{-1}^{+1} \frac{dz}{\sqrt{1-z^2}}.$$

We have seen (§ 43) that we can determine the value of this integral by regarding the closed line as a line which encloses the point ∞, and consequently we obtain

$$A = -2\pi.$$

For a line running in the second sheet we should have obtained the value $+2\pi$; and, in fact, a line which makes a circuit round -1 and $+1$ in the second sheet in the direction of increasing angles crosses the cross-cut in a direction opposite to that of a similar line in the first sheet. Hence the inverse function $\sin w$ of the preceding integral is periodic with the period 2π.

In order to determine the mode of representing the z-surface on the w-surface, we will let z describe the entire boundary of T' in the positive direction, beginning at a, where w has a value denoted by w_a. If the outer boundary situated in the first sheet be described by the variable z, then w goes from w_a to $w_a - 2\pi$ along a line the form of which depends upon the form of the boundary-line in z (Fig. 64). Now let z go from a to c along the left edge (directed from a to c) of the cross-cut ac, and w from $w_a - 2\pi$ to a value which may be denoted by w_c. The line along which w moves may again differ in form according to the form of the cross-cut ac. Let z next describe from c the outer boundary of the second sheet; then w goes from w_c to $w_c + 2\pi$ along a curve which depends only upon the outer boundary of the second sheet of the z-surface. Finally z closes its circuit, by returning along the left edge (directed from c to a) of the cross-cut ca to the initial point; then w also returns from $w_c + 2\pi$ to w_a. The line along which w last moves must be parallel to the path $(w_a - 2\pi, w_c)$, because these two lines correspond to the two edges of the cross-cut, and because w has values which differ by 2π at every pair of infinitely near points on the two edges. If we now enlarge indefinitely the outer boundaries of the surface T, then the lines $(w_a, w_a - 2\pi)$ and $(w_c, w_c + 2\pi)$ move away to infinity, and z, or $\sin w$, acquires all its values in one strip, which is bounded by the parallel lines AB and CD. But in such a strip z acquires all its values twice; for, since the z-surface consists of two sheets, there correspond two values of w to each value of z, not taking into consideration the modulus of periodicity, and hence z, or $\sin w$, acquires the same value at two different points w.

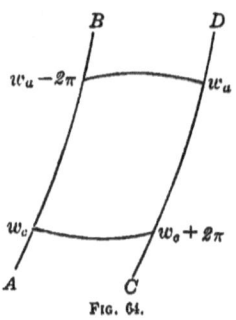

Fig. 64.

If we assume that the cross-cut ac runs along the ordinate axis, so that on both its edges $z = iy$ (where y is real), we obtain

$$w = i \int_0^y \frac{dy}{\sqrt{1+y^2}};$$

thus w is also a pure imaginary or differs from a pure imaginary quantity by multiples of the real modulus of periodicity 2π. The w-plane is then divided into strips by parallel straight lines, which run parallel to the y-axis and pass through the points $0, \pm 2\pi, \pm 4\pi$, etc.

In order to determine the relation between two points w and w' in the same strip, to which correspond equal values of z, we let the latter variable first pass from the point 0 in the first sheet to the point $0'$ in the second sheet, situated immediately below 0, without crossing the cross-cut. This is done (Fig. 65) by passing along the branch-cut round $+1$, next along the other side of the same and then across the branch-cut into the second sheet. On this path we obtain at $0'$ the value

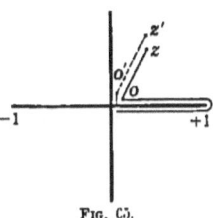

Fig. 65.

$$w = \int_0^1 \frac{dz}{\sqrt{1-z^2}} - \int_1^0 \frac{dz}{\sqrt{1-z^2}} = \pi.$$

Consequently, the point $w = \pi$ corresponds to the point $z = 0'$ situated in the second sheet. If z now go from 0 to z in the first z-sheet, w goes from 0 to w. But if z go in the second sheet from $0'$ to z', where z' is situated immediately below z, then w starts with the value π, and because the radical $\sqrt{1-z^2}$ has the negative sign in this part, it acquires at z' the value

$$w' = \pi - \int_0^z \frac{dz}{\sqrt{1-z^2}};$$

but
$$w = \int_0^z \frac{dz}{\sqrt{1-z^2}};$$

and consequently
$$w + w' = \pi,$$

or the sum of the two values of w, for which z, or $\sin w$, acquires the same value, is equal to half the modulus of periodicity, not taking into consideration multiples of the latter.

4. *The Elliptic Integral.*

$$w = \int_0^z \frac{dz}{\sqrt{(1-z^2)(1-k^2z^2)}}.$$

Here the z-surface consists likewise of two sheets, and has the four discontinuity- and branch-points $+1$, -1, $+\frac{1}{k}$, $-\frac{1}{k}$. None of these points need be excluded, because the function under the integral sign becomes infinite at each of them only of the order $\frac{1}{2}$. The point ∞ also need not be excluded, since

$$\lim [zf(z)]_{z=\infty} = \lim \left[\frac{z}{\sqrt{(1-z^2)(1-k^2z^2)}} \right]_{z=\infty} = 0.$$

Consequently, in this case no point need be excluded. This is in conformity with the condition that the preceding integral, as we have already seen (§ 45), remains finite for every value of z, and hence can become infinite only by the addition of an infinitely great multiple of a modulus of periodicity. If we assume that the z-surface is closed at infinity, we have to do with a surface which is not bounded at all (or only by an arbitrary point), but which is multiply connected. In such a surface we let the first cross-cut be a line returning into itself (§ 47). If we assume that the points -1 and $+1$ on the one hand, and $+\frac{1}{k}$ and $-\frac{1}{k}$ on the other, are connected by branch-cuts,[1] we will take for the first cross-cut a line q_1, which encloses the two points -1 and $+1$ in the upper sheet (Fig. 66). Such a line does not divide the surface, since we can pass from one side to the other side of the same. The way in which this passage is made (cf. § 46, v.) indicates how the second cross-cut q_2 is to be drawn; namely, from

---

[1] In Fig. 66 it has been likewise assumed that k is real and less than unity; then the branch-cut drawn from $+\frac{1}{k}$ to $-\frac{1}{k}$ passes through ∞. But we will first consider k as a quite arbitrary quantity, and only later return to the assumption that k is real and less than unity.

MODULI OF PERIODICITY. 271

a point a of the first cross-cut across the branch-cut $(-1, +1)$ into the second sheet, then across the other branch-cut back again into the first sheet, returning in this sheet to the initial point, but on the other side of the first cross-cut (to a'''). These two lines now form together a continuous path, in which each

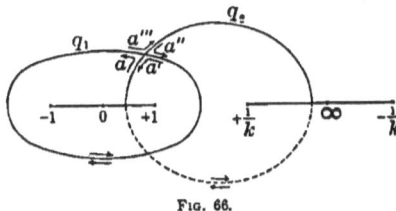

FIG. 66.

of the two cross-cuts is described twice in opposite directions. The arrows indicate this description in the positive direction. In this surface T' every closed line forms by itself alone the complete boundary of a portion of the surface, and hence the surface is simply connected. Its boundary is formed by the two edges of the cross-cuts. Thus the original surface was triply connected.

The modulus of periodicity A_1 for the cross-cut q_1 is the integral $\int dw$, taken in the direction of increasing angles along a closed line which leads from one side of the cross-cut to the other side of the same, e.g., along q_2. This line can be contracted until it coincides with two straight lines, one of which leads from $\frac{1}{k}$ to 1 in the first sheet, the other from 1 to $\frac{1}{k}$ in the second sheet. If we then assume that the sign $+$ is to be attached to the radical in the first sheet, and if for brevity we let

$$\sqrt{(1-z^2)(1-k^2z^2)} = \Delta(z, k),$$

we have

$$A_1 = \int_{\frac{1}{k}}^{1} \frac{dz}{\Delta(z, k)} - \int_{1}^{\frac{1}{k}} \frac{dz}{\Delta(z, k)} = -2\int_{1}^{\frac{1}{k}} \frac{dz}{\Delta(z, k)}.$$

The modulus of periodicity A_2 for the second cross-cut is in like manner equal to the integral taken along the line q_1 and this line, as in the former case, can be contracted up to the branch-cut; then, as before,

$$A_2 = \int_{+1}^{-1} \frac{dz}{\Delta(z, k)} - \int_{-1}^{+1} \frac{dz}{\Delta(z, k)} = -2 \int_{-1}^{+1} \frac{dz}{\Delta(z, k)},$$

or also, as is evident,

$$A_2 = -4 \int_0^1 \frac{dz}{\Delta(z, k)}.$$

The elliptic integral therefore has two different moduli of periodicity; consequently the inverse function, the so-called *elliptic function*, which is designated after Jacobi by sin am w, is doubly periodic.

If we now represent the z-surface on the w-surface, we obtain the following results: If z go from a along the cross-cut q_1 in the direction of increasing angles and at the same time in the positive boundary-direction, and therefore return to a on the inner edge of the line q_1 (in Fig. 66 from a to a'), then w increases from w to $w + A_2$. In this w passes along a line (Fig. 67) the form of which depends upon the form of the line q_1 (to be chosen arbitrarily); if z next go along the line q_2 in the same direction to a again (i.e., from a' to a''), w increases from $w + A_2$ to $w + A_1 + A_2$, along a line which changes its form with that of q_2. If z then describe the line q_1 starting from a'', always in the positive boundary-direction, but now in the direction of decreasing angles (i.e., from a'' to a'''), w goes from $w + A_1 + A_2$ to $w + A_1$, because it is diminished by A_2. The line along which this movement of w takes place must be parallel to the line $(w, w + A_2)$, because the two values of w at every two infinitely near points on the two edges

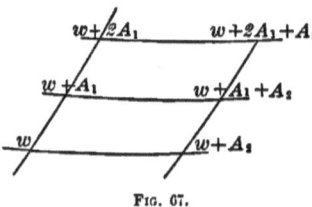

Fig. 67.

of the line q_1 differ by the quantity A_1, and hence two different but parallel lines in w correspond to the two edges of this cross-cut. Finally, if z go from a''' to a along the cross-cut q_2, then w goes from $w + A_1$ to w along a line which for the same reason as before must be parallel to the line $(w + A_2, w + A_1 + A_2)$. Thus to the two edges of the cross-cut q_1 correspond the parallel lines $(w, w+A_2)$ and $(w + A_1, w+A_1+A_2)$, and to the two edges of the cross-cut q_2 the parallel lines $(w, w + A_1)$ and $(w + A_2, w + A_1 + A_2)$. Now to all the points z in the whole infinite extent of the z-surface correspond only such points w as lie within[1] the curvilinear bounded parallelogram, for a line can be drawn through any arbitrary point of the z-surface which leads from one side of q_1 to the other side of q_1, without crossing a cross-cut; hence the corresponding line w leads from the line $(w, w + A_2)$ through the interior of the parallelogram to the line $(w + A_1, w + A_1 + A_2)$. Consequently z, or sin am w, acquires all its values in this parallelogram, and indeed each value twice, since the z-surface consists of two sheets.

Other parallelograms now adjoin this parallelogram on all sides. For if we let z go from a to a''', for instance, then w goes from w to $w + A_1$. But if we now let w proceed continuously across the cross-cut q_1, then w starts with the value $w+A_1$; hence to the side $(w + A_1, w + A_1 + A_2)$ is joined a new parallelogram, at the corners of which w has the values

$$w + A_1, w + A_1 + A_2, w + 2A_1 + A_2, w + 2A_1.$$

Similarly for the three other sides. In this way the whole w-plane is divided into parallelograms by two sets of parallel lines. If we assume that k is real and less than unity, the four points $+1, -1, +\dfrac{1}{k}, -\dfrac{1}{k}$ lie on the principal axis; if we now contract the two cross-cuts, so that they run along the two edges of the principal axis, the parallel lines become straight lines, which run parallel to the x- and the y-axis respectively.

[1] Within, because w remains finite for all values of z.

In this case, then, $\int_0^1 \dfrac{dz}{\sqrt{(1-z^2)(1-k^2z^2)}}$

is real. We usually designate the value of this integral after Jacobi by K. The other integral

$$\int_1^{\frac{1}{k}} \dfrac{dz}{\sqrt{(1-z^2)(1-k^2z^2)}},$$

on the other hand, is a pure imaginary. If we let $\sqrt{1-k^2} = k'$ and transform the integral by the substitution

$$z = \dfrac{\sqrt{1-k'^2 z'^2}}{k},$$

we get $\qquad -i\int_0^1 \dfrac{dz'}{\sqrt{(1-z'^2)(1-k'^2 z'^2)}},$

which is designated by $-iK'$. Consequently the moduli of periodicity, except as to signs, are

$$4K \text{ and } 2iK'.$$

We can in this example also determine the relation between every pair of values of w which correspond to the same value of z; i.e., to two points of the z-surface lying one immediately below the other. To the value $z = 0$ in the first sheet corresponds $w = 0$. In order to come to $0'$ in the second sheet, we can conceive the cross-cut q_2 to be so enlarged that it also encloses the origin as well as the points 1 and $\dfrac{1}{k}$. We can pass within T' from 0 along the branch-cut round the point $+1$ to the other side of the branch-cut and then across the same to $0'$ (cf. p. 269); w then acquires at $0'$ the value

$$\int_0^1 \dfrac{dz}{\Delta(z,k)} - \int_1^0 \dfrac{dz}{\Delta(z,k)} = 2K,$$

and is therefore equal to the half of one of the moduli of periodicity. If z now go from $0'$ to z', where z' lies in the

MODULI OF PERIODICITY.   275

second sheet immediately below z, we have, designating the value of w at z' by w',

$$w' = 2K - \int_0^z \frac{dz}{\Delta(z, k)};$$

but
$$w = \int_0^z \frac{dz}{\Delta(z, k)},$$

and, therefore,
$$w + w' = 2K.$$

If we take the integral $\int dw$ along a closed line which encloses all four branch-points, such a line runs entirely in the first sheet (Fig. 66), and hence forms by itself alone a com-

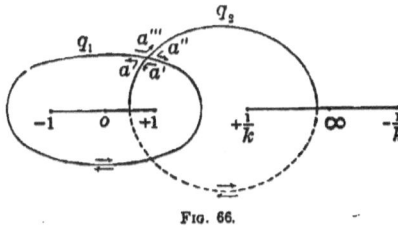

Fig. 66.

plete boundary. Consequently, this integral has the value zero. If we now contract this line up to the principal axis, on which are the four branch-points, the integral is divided into the following parts (the lines may be described in the direction of decreasing angles):

(1) from -1 to $+1$;
(2) from $+1$ to $+\frac{1}{k}$;
(3) from $+\frac{1}{k}$ through ∞ to $-\frac{1}{k}$;
(4) from $-\frac{1}{k}$ through ∞ to $+\frac{1}{k}$;
(5) from $+\frac{1}{k}$ to $+1$;
(6) from $+1$ to -1.

The radical is to be taken negatively in (6) and (4), because for these the path of integration lies on the right side of the branch-cuts $(-1, +1)$ and $\left(+\frac{1}{k}, -\frac{1}{k}\right)$; in all the others it is

276   THEORY OF FUNCTIONS.

to be taken positively. Consequently (2) and (5) cancel each other, and (1) and (3) are to be doubled. Since, further,

$$(1) = 2\int_0^1 \frac{dz}{\Delta(z, k)}, \quad (3) = 2\int_{\frac{1}{k}}^{\infty} \frac{dz}{\Delta(z, k)},$$

we obtain
$$\int_0^1 \frac{dz}{\Delta(z, k)} + \int_{\frac{1}{k}}^{\infty} \frac{dz}{\Delta(z, k)} = 0,$$

and hence
$$\int_{\frac{1}{k}}^{\infty} \frac{dz}{\Delta(z, k)} = -K.$$

From this result follows also the value of the integral between the limits 0 and ∞; for since this is divided into the parts $0 \cdots 1$, $1 \cdots \frac{1}{k}$, $\frac{1}{k} \cdots \infty$, we obtain

$$\int_0^{\infty} \frac{dz}{\Delta(z, k)} = K - iK' - K = -iK',$$

or, since we can add to this value the modulus of periodicity $2iK'$, also

$$\int_0^{\infty} \frac{dz}{\Delta(z, k)} = iK'.$$

Thus z becomes infinite within the parallelogram with the corners 0, $4K$, $4K + 2iK'$, and $2iK'$ for $w = iK'$ and $w = 2K + iK'$.

We will also in this example, following the method of Riemann, consider the relation between the doubly periodic function and the elliptic integral in the inverse manner, i.e., starting from the doubly periodic function. Let $\phi(w)$ be a uniform doubly periodic function, and therefore possess the property that simultaneously

$$\phi(w + A_1) = \phi(w) \text{ and } \phi(w + A_2) = \phi(w).$$

Then the straight lines which represent the complex quantities A_1 and A_2 must have different directions. For if they have the same direction, A_1 and A_2 must possess a real ratio (§ 2, 3). This can be either rational or irrational. If it be rational,

A_1 and A_2 are commensurable, and hence are multiples of one and the same quantity B. We can thus let

$$A_1 = mB, \quad A_2 = nB,$$

wherein m and n denote two integers, which are relatively prime to each other, and we then obtain

$$\phi(w) = \phi(w + mB) = \phi(w + nB).$$

Now since in this case there are two integers a and b connected by the relation
$$ma - nb = 1,$$
and since, moreover,
$$\phi(w + maB - nbB) = \phi(w),$$
we also have $\quad \phi(w + B) = \phi(w),$

and hence in this case the function $\phi(w)$ is simply and not doubly periodic. But if A_1 and A_2 have a real irrational ratio, so that they are incommensurable, there are always two integers m and n for which the modulus of $mA_1 + nA_2$ becomes less than any assignable quantity.[1] Since now also

$$\phi(w + mA_1 + nA_2) = \phi(w),$$

[1] If we let $\dfrac{A_2}{A_1} = a$, then, according to the assumption, a is real and irrational. If we develop the absolute value $|a|$ of a in a continued fraction, and if we denote two consecutive convergents of the same by $\dfrac{\mu}{\nu}$ and $\dfrac{\mu'}{\nu'}$, then, as is well known, for the absolute value

$$\left\{\frac{\mu}{\nu} - |a|\right\} < \frac{1}{\nu\nu'},$$

and hence $\quad (\mu - \nu|a|) < \dfrac{1}{\nu'}.$

But since the denominator of the convergents increases indefinitely, we can make this expression as small as we please by continuing the development sufficiently far. But we have

$$mA_1 + nA_2 = A_1(m + na);$$

hence if we let $m = \mu$ and $n = \mp \nu$, according as $|a| = \pm a$, we can make $m + na$, and therefore also the modulus of $mA_1 + nA_2$, as small as we please.

the function $\phi(w)$ maintains the same value for an indefinitely small change of the variable, and hence is a constant. Consequently the ratio of the two moduli of periodicity of a doubly periodic function must be imaginary, and therefore the straight lines A_1 and A_2 must have different directions. Then we can divide the w-plane into parallelograms by two sets of parallel lines in such a way that $\phi(w)$ acquires the same values on any two parallel lines; moreover, it then acquires all its values in each parallelogram, and has the same value at every two corresponding points of different parallelograms.

Since the uniform function $\phi(z)$ must become infinite for some one value of w (§ 28), it must become infinite in every parallelogram. Let us, therefore, select any parallelogram (Fig. 68), and let r, r', r'', etc., be the points of the same at which $\phi(w)$ becomes infinite. If we form the integral

$$\int \phi(w)dw,$$

taken over the boundary of the parallelogram, then by § 19 this is equal to the sum of the integrals taken round the points of discontinuity r, r', r'', etc. Therefore, if $\phi(w)$ at these points become infinite in the same way as

$$\frac{c}{w-r}+\cdots,\quad \frac{c'}{w-r'}+\cdots,\quad \frac{c''}{w-r''}+\cdots,\text{ etc.,}$$

respectively do, we have

$$\int \phi(w)dw = 2\pi i(c+c'+c''+\cdots).$$

But $\phi(w)$ has the same values on the side CE as on DF, the same values on CD as on EF, and in the description of the boundary of the parallelogram the parallel sides are described in opposite directions; hence the integrals taken along these sides cancel each other, and thus

$$\int \phi(w)dw = 0,$$

consequently, also $\quad c+c'+c''+\cdots=0.$

MODULI OF PERIODICITY. 279

From this we conclude that $\phi(w)$ must become infinite more than once in each parallelogram, and at least infinite of the first order at two points or of the second order at one point. If n denote the multiplicity of the infinite value (or values) of $\phi(w)$ in each parallelogram, we can first show that $\phi(w)$ must acquire each value h in each parallelogram n times. For that purpose we will consider the integral

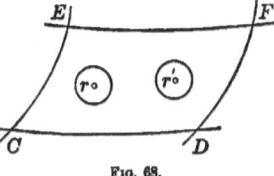

Fig. 68.

$$\int d \log [\phi(w) - h] \text{ or } \int \frac{\phi'(w) dw}{\phi(w) - h},$$

taken along the boundary of the parallelogram. This also has the value zero, because both $\phi(w) - h$ and $\phi'(w)$ have the same values on the opposite sides of the parallelogram. But on the other hand this integral is equal to the sum of the integrals taken round those points at which $\phi'(w)$ becomes infinite, and round those at which $\phi(w) - h$ vanishes. The former are the same as those at which $\phi(w)$ or $\phi(w) - h$ becomes infinite (§ 29). Now if in general a be a point at which $\phi(w) - h$ becomes either infinitesimal or infinite, and that of the pth order (p positive for infinitesimal values), we can put (§ 34)

$$\phi(w) - h = (w - a)^p \psi(w),$$

wherein $\psi(w)$, for $w = a$, is neither zero nor infinite. We then obtain

$$\int d \log [\phi(w) - h] = p \int \frac{dw}{w - a} + \int \frac{\psi'(w) dw}{\psi(w)} = 2\pi i p.$$

Therefore $\quad \int d \log [\phi(w) - h] = 2\pi i \Sigma p,$

taken round the entire parallelogram, and hence

$$\Sigma p = 0.$$

Now $\phi(w) - h$ becomes n times infinite, just as $\phi(w)$ does; if m denote the number of times that it becomes zero, we have

$$\Sigma p = m - n = 0,$$

and hence $m = n$.

Since, therefore, $\phi(w) - h$ must become zero n times, $\phi(w)$ also becomes n times equal to h.

We will now consider in the following only the simplest case, in which $\phi(w)$ becomes infinite twice in each parallelogram and therefore also acquires every value twice. We will first assume that $\phi(w)$ becomes infinite of the first order at two points r and s. Then, denoting $\phi(w)$ by z, we can put

$$z = \phi(w) = \frac{c}{w-r} + \frac{c'}{w-s} + \psi(w),$$

or, since $c + c' = 0$,

$$z = \phi(w) = \frac{c}{w-r} - \frac{c}{w-s} + \psi(w), \qquad (4)$$

wherein c denotes a given constant, and $\psi(w)$ a function which no longer becomes infinite in the parallelogram under consideration, and therefore only in the other parallelograms at the points $r + mA_1 + nA_2$ and $s + mA_1 + nA_2$ (wherein m and n are to have all positive and negative integral values). We will first determine the relation between the two values of w for which $\phi(w)$ has the same value. For this purpose let

$$v = r + s - w.$$

If we substitute v for w in (4), we get

$$\phi(v) = \frac{c}{v-r} - \frac{c}{v-s} + \psi(v).$$

But since
$$v - r = -(w-s)$$
$$v - s = -(w-r),$$

it follows that $\quad \phi(v) = -\dfrac{c}{w-s} + \dfrac{c}{w-r} + \psi(v),$

and hence $\quad \phi(w) - \phi(v) = \psi(w) - \psi(v).$

Therefore this difference remains finite in the first parallelogram. In an adjacent parallelogram $\phi(w)$ becomes infinite at $w=r+A_1$ and $w = s + A_1$; hence we can also let

$$\phi(w) = \frac{c_1}{w - r - A_1} - \frac{c_1}{w - s - A_1} + \psi_1(w),$$

wherein now $\psi_1(w)$ remains finite for all points of the second parallelogram. If we now substitute

$$v_1 = r + s + 2A_1 - w,$$

we get $\qquad w - r - A_1 = -(v_1 - s - A_1)$

$$w - s - A_1 = -(v_1 - r - A_1),$$

and hence also

$$\phi(v_1) = -\frac{c_1}{w - s - A_1} + \frac{c_1}{w - r - A_1} + \psi_1(v_1);$$

consequently $\qquad \phi(w) - \phi(v_1) = \psi_1(w) - \psi_1(v_1)$

and remains finite within the second parallelogram. But since v_1 differs from v only by twice the modulus of periodicity A_1, it follows that

$$\phi(v_1) = \phi(v), \ \phi(w) - \phi(v_1) = \phi(w) - \phi(v);$$

hence the difference $\qquad \phi(w) - \phi(v)$

remains finite in the second as well as in the first parallelogram. If we continue in this way from parallelogram to parallelogram, we conclude that this difference does not become infinite in any parallelogram and hence not at all; therefore it must be a constant. To find the value of this constant, we let

$$w = \frac{r + s}{2};$$

then $\qquad v = \frac{r + s}{2} = w,$

and since the function ϕ is uniform, also

$$\phi(v) = \phi(w).$$

Therefore, since the difference $\phi(w) - \phi(v)$ has the value zero for one value of w, it has this value always, and hence
$$\phi(r + s - w) = \phi(w).$$
Consequently w and $r + s - w$ are the two corresponding values of w for which the function $\phi(w)$ acquires the same value.

From (4) it follows that
$$\phi'(w) = \frac{dz}{dw} = -\frac{c}{(w-r)^2} + \frac{c}{(w-s)^2} + \psi'(w);$$
therefore, not taking into account the moduli of periodicity, the derivative $\phi'(w)$ is infinite only for $w = r$ and $w = s$, but for these it is infinite of the second order. Hence it becomes infinite four times in every parallelogram and therefore also acquires each value four times. It is likewise a uniform function of w; but it is important to inquire whether it is also a uniform function of z. Now the derivative $\frac{dz}{dw}$ acquires the same value at every pair of corresponding points of different parallelograms at which z has the same value. Thus we have to consider only the points v and w of the same parallelogram. If we differentiate the equation
$$\phi(w) = \phi(v)$$
as to w, we obtain $\quad \phi'(w) = -\phi'(v),$

since $\quad\quad\quad\quad \dfrac{dv}{dw} = -1.$

Consequently z does indeed take the same value for v and u, but $\frac{dz}{dw}$ opposite values; therefore $\frac{dz}{dw}$ is not a uniform function of z, since it can acquire two different values for the same value of z. But since these are numerically equal and of opposite signs, it follows that $\left(\dfrac{dz}{dw}\right)^2$ is a uniform function of z. Now $\frac{dz}{dw}$ is infinite only where z is also infinite, but it is infinite of the second order while z is infinite of the first order; consequently $\left(\dfrac{dz}{dw}\right)^2$ is infinite of the fourth order. Therefore $\left(\dfrac{dz}{dw}\right)^2$ is a uniform function of z, which becomes infinite only for $z = \infty$ and that of the fourth order; accordingly it is an integral function of the fourth degree. Such a function is also four times zero. If we denote by α, β, γ, δ, the values of z for which it becomes zero, and by C a constant, we have

(5) $$\left(\dfrac{dz}{dw}\right)^2 = C(z-\alpha)(z-\beta)(z-\gamma)(z-\delta);$$

from this is obtained

$$w = \int \dfrac{dz}{\sqrt{C(z-\alpha)(z-\beta)(z-\gamma)(z-\delta)}}.$$

Hence a doubly periodic function which becomes twice infinite of the first order in every parallelogram, is the inverse function of an elliptic integral. The constant C can be expressed in terms of c. For since by (5)

$$C = \lim \left[\dfrac{\left(\dfrac{dz}{dw}\right)^2}{z^4}\right]_{z=\infty},$$

we obtain

$$C = \lim \left\{\dfrac{\left[-\dfrac{c}{(w-r)^2} + \dfrac{c}{(w-s)^2} + \psi'(w)\right]^2}{\left[\dfrac{c}{w-r} - \dfrac{c}{w-s} + \psi(w)\right]^4}\right\}_{w=r}$$

$$= \lim \left\{\dfrac{\left[-c + \dfrac{c(w-r)^2}{(w-s)^2} + (w-r)^2\psi'(w)\right]^2}{\left[c - \dfrac{c(w-r)}{w-s} + (w-r)\psi(w)\right]^4}\right\}_{w=r}$$

$$= \dfrac{c^2}{c^4} = \dfrac{1}{c^2}.$$

Then $$w = \int \dfrac{c\,dz}{\sqrt{(z-\alpha)(z-\beta)(z-\gamma)(z-\delta)}}$$

This integral admits of the same treatment as the former

$$\int \frac{dz}{\sqrt{(1-z^2)(1-k^2z^2)}},$$

if we put the four branch-points, α, β, γ, δ, in place of $+1$, -1, $+\frac{1}{k}$, $-\frac{1}{k}$; it can also be transformed into the latter.

We now proceed to the case in which the function $\phi(w)$ becomes infinite only at one point, but of the second order. In this case we must put

$$z = \phi(w) = \frac{c}{(w-r)^2} + \psi(w);$$

for the term containing $(w-r)^{-1}$ must be wanting in order that $\int \phi(w) dw$, extended over the boundary of the parallelogram, may have the value zero. We infer in this case, just as before, that

$$\phi(2r-w) = \phi(w),$$

by letting $s = r$, and hence

$$\phi'(2r-w) = -\phi'(w).$$

Therefore $\frac{dz}{dw}$ is not a uniform function of z, but again $\left(\frac{dz}{dw}\right)^2$ is a uniform function of z. In this case

$$\frac{dz}{dw} = -\frac{2c}{(w-r)^3} + \psi'(w);$$

thus $\frac{dz}{dw}$ becomes infinite, of the third order, only where z is infinite of the second order. Therefore $\frac{dz}{dw}$, as a function of z, is infinite of the order $\frac{3}{2}$ for $z = \infty$, and consequently $\left(\frac{dz}{dw}\right)^2$ is infinite of the third order. Accordingly in this case we have

$$\left(\frac{dz}{dw}\right)^2 = C(z-\alpha)(z-\beta)(z-\gamma).$$

Therein
$$C = \lim \left[\frac{\left(\frac{dz}{dw}\right)^2}{z^3} \right]_{z \doteq \infty} = \lim \left\{ \frac{\left[-\frac{2c}{(w-r)^2} + \psi'(w) \right]^2}{\left[\frac{c}{(w-r)^2} + \psi(w) \right]^3} \right\}_{w \doteq r}$$

$$= \lim \left\{ \frac{[-2c + (w-r)^3 \psi'(w)]^2}{[c + (w-r)^2 \psi(w)]^3} \right\}_{w \doteq r} = \frac{4c^2}{c^3} = \frac{4}{c};$$

consequently
$$\left(\frac{dz}{dw}\right)^2 = \frac{4}{c}(z - \alpha)(z - \beta)(z - \gamma),$$

and
$$w = \tfrac{1}{2} \int \frac{\sqrt{c}\, dz}{\sqrt{(z - \alpha)(z - \beta)(z - \gamma)}},$$

which is likewise an elliptic integral.

We here close this discussion, because it is not the purpose of this book to enter more in detail into the investigation of periodic functions; but the cases treated are to be regarded only as examples illustrating the general considerations.

Supplementary Note to Riemann's Fundamental Proposition on Multiply Connected Surfaces.

Riemann originally gave to the proposition bearing his name (§ 49) a somewhat different and more general enunciation, which presents many advantages, while it removes at once a difficulty which otherwise requires supplementary examination.

This differs from the form of the proposition as enunciated in § 49 in the following manner: If the surface T be first modified by q_1 cross-cuts of a first mode of resolution into a system T_1, which consists of a_1 pieces, and a second time by q_2 cross-cuts of a second mode of resolution into a system T_2, which consists of a_2 pieces, then in contradistinction to the enunciation of § 49 it is only assumed that the a_1 pieces of the system T_1 are all simply connected, while the a_2 pieces of the system T_2 may be arbitrary; then the property that $q_2 - a_2$ cannot be greater than $q_1 - a_1$ holds, and therefore

$$q_2 - a_2 \lesseqgtr q_1 - a_1.$$

In the proof of this property, the first main division of the proof remains exactly the same as in § 49 or § 51. By the superposition of the two systems of cross-cuts a new system of surfaces $\mathfrak{T}$ is produced in two ways, and it is proved that if the lines of the second mode of resolution, when drawn in T_1, form $q_2 + m$ cross-cuts in that surface, then the lines of the first mode of resolution, when drawn in T_2, also consist of $q_1 + m$ cross-cuts. Since, moreover, T_1, according to the hypothesis, consists of a_1 simply connected pieces, therefore $\mathfrak{T}$ consists of
$$\mathfrak{A} = a_1 + q_2 + m$$
pieces.

Now the system $\mathfrak{T}$ is also derived from T_2, which consists of a_2 pieces, by $q_1 + m$ cross-cuts. Therefore the number $\mathfrak{A}$ of pieces of which $\mathfrak{T}$ consists can (by § 48, V., note) be not greater than $a_2 + q_1 + m$, but on the contrary

$$\mathfrak{A} \gtrless a_2 + q_1 + m,$$

i.e., $$a_1 + q_2 + m \gtrless a_2 + q_1 + m;$$

from this follows immediately

$$q_2 - a_2 \gtrless q_1 - a_1,$$

which was to be proved.

Therefore $q_2 - a_2$ cannot be greater than $q_1 - a_1$; and if the case occur, that the numbers a_1 and a_2 of pieces arising from the two modes of resolution are equal to each other, then q_2 cannot be greater than q_1.

Conversely, if T_2 consist of only simply connected pieces (in number a_2), while the a_1 pieces which form the system T_1 are arbitrary, we have

$$q_1 - a_1 \gtrless q_2 - a_2.$$

But if both systems T_1 and T_2 consist of only simply connected pieces, then $q_2 - a_2$ cannot be greater than $q_1 - a_1$, nor $q_1 - a_1$ be greater than $q_2 - a_2$; hence in this case

$$q_1 - a_1 = q_2 - a_2,$$

and this is the principle of § 49.

From the above form of Riemann's fundamental proposition is at once derived the second proposition of § 52, upon which the classification of surfaces depends. It is here assumed that a multiply connected surface T can be changed into a simply connected surface T_1 by q cross-cuts drawn in a definite manner, and it will be proved that this modification is always effected by means of q non-dividing cross-cuts, in whatever way also the latter may be drawn. From the above proposition follows, first, that the surface T cannot be made simply connected by less than q cross-cuts; hence by § 48, II., it is possible to draw q cross-cuts in such a definite way that T is likewise not divided. Then a surface T_2, which consists of a single piece, again arises. But this cannot be multiply connected; for if it were, we could still draw in it at least one non-dividing cross-cut (§ 48, II.) and thus obtain by $q+1$ cross-cuts a surface consisting of one piece, while the simply connected surface T_1 arose through q cross-cuts; but this contradicts the above proposition in the original Riemann form.

The property proved in § 53, V., also requires no further proof if this form of the proposition serve as the basis, but follows at once. The question here is concerning a $(q+1)$-ply connected surface T, which is therefore made simply connected by q cross-cuts and is divided by one additional cross-cut into two pieces. If a dividing cross-cut R be first drawn instead of these, by which T is divided into two pieces A and B, and if in these pieces additional cross-cuts be drawn, we still have two pieces, if neither A nor B be divided by the new cross-cuts. But then the number of these new cross-cuts possible in A and B cannot, according to our proposition, be greater than q, and is therefore a finite number; from this the remainder follows, as in § 53, V.

www.ingramcontent.com/pod-product-compliance
Lightning Source LLC
Chambersburg PA
CBHW031248250426
43672CB00029BA/1376